ISBN-13: 978-0-6459423-5-4

Cover design by: Good Beach Reads

EARTH TO CLIMATE CHANGE

Contents

PART 1 - CHRONOLOGY OF THE ANCESTORS

"A great wonder appeared in Gaul at Fort Tauredunum, located on the river Rhone on the side of a mountain. For more than sixty days an indescribable roar was heard, then eventually a mountain fractured and separated from the next, collapsing into the river, carrying people, churches, property and houses. The river course became blocked and the water flowed backwards, flooding the area upstream, inundating and destroying everything on the banks". History of the Gauls, Gregoire de Tours, 563.

THE TAUREDUNUM EVENT

On the Celtic road network, underneath Les Dents du Midi, a late Iron Age Oppidum, the chef-lieu of the Nantuaten, Taranais Dunum, probably had its fortifications improved between the first attempt and the successful Roman invasion in 15 BCE. Buried in the pre-563 moraines, four Bronze Age bracelets were found at the level of the identified site of the settlement before the landslide destroyed it. By the sixth century the Romanized name was Tauredunum or Tauredun. This fortress was built just above the river plain, where there was once a low but defensible plateau under the towering, majestic Dents du Midi. That it was the original site of the chef-lieu of the Nantuaten, Taranais, is well borne out by archaeological finds. It must have grown over the centuries, and it was located just above Massongex, a couple of kilometres from the current site of Saint Maurice, where the Romans had preferred to develop their administrative centre, Agaunum. The location is on the edge of the old Nantuaten territory boundary, an excellent location to control trade through the narrow corridor sandwiched between Les Dents du Midi and Dent de Morcles.

Celtic toponymy, the study of place names wholly or partially of Celtic origin identifies 'dunum' as a derivation of the Celtic dunon or 'fortress'. Europe is scattered with places whose second part is the Romanized 'dunum'. For example, Lugdunum was the fortress of Lug, the God of Light. Tarnaias was a Nantuaten stronghold, and 'dunum' may refer to the fortifications built to counter the threat of Roman invasion. A strategic location, it held out against the original Roman army invasion for a negotiated settlement. Probably Tarnaias was Taranis, in Celtic mythology, the god of thunder, essentially worshipped in Gaul. This would make sense as a place name, as during the hot summer months, violent thunderstorms can rage around the surrounding mountains. As geological detective work has placed the course of the Rhone much closer to Les Dents du Midi before the seismic event, it is likely that the settlement was strategically located on the hillside above the river that flowed below it.

The fall of the mountain, dislodging and burying not only the fortress and its surrounding dwellings, but also all the nearby habitations in the year 563, was an event of momentous proportions. It caused a tidal wave that eventually swamped Geneva, engulfing the settlements on both banks of Lake Leman. The Tauredunum event was of such magnitude that it was recorded not once, but twice, independently and separately, by the two western European monk historians of the age. Tarnaias was on the south side of the Rhone. One of the pieces of the jigsaw for the original location of Tarnaias (which by 563 became Tarnaias Dunum, or Tauredunum) was the finding of the remains of the limestone blocks that collapsed at the same time as the mountain above it. The blocks formed the abutments of the Roman bridge across the Rhone, and were discovered in 1976 in the bed of the Rhone at Massongex, by a mechanical digger extracting gravel. They would have been deposited there by the unleashing of the tidal wave, following the landslide of Les Dents du Midi. The bridge, in 563, would have crossed the Rhone at a strategic place where the river was narrow, and the ground firm enough to support the monumental structure of stone. The fortress was thus located close to the narrow passage of the Rhone adjacent to the negotiated boundary between the territory of the Nantuaten and the Veragri, whose tribal capital at Octodurum, was a mere fifteen kilometres or so further upstream along the Rhone.

Saint Maurice was originally called Agaunum, and the name does come from the Gaulish language, originally meaning 'stone'. The stone may refer to the large quantity of terminal moraine from the glacier deposited in the bed of the Rhone at this point, much of it swept away in 563. For both the Nantuaten and the Romans, it allowed the regulation of the travellers along the trade route over the Alps, and may well have been a lucrative source of revenue, perhaps an early customs post demanding duty to be paid on transported goods. The collapse must have been triggered by a major seismic event, as the adjoining mountain Grammont also suffered a massive rock avalanche in the same event. No doubt the whole plain must have been violently shaken. Yet not everyone heeded the warnings, even though the preliminary earthquake tremors made themselves heard preceding the landslides. The event was recorded in some detail by Gregory of Tours in his History of the Franks. He wrote:

'Le Château de Tauredun: A great prodigy appeared in Gaul at the fortress of Tauredunum, which was situated on high ground above the River Rhone. Here a curious bellowing sound was heard for more than sixty days: then the whole hillside was split open and separated from the mountain nearest to it, and it fell into the river, carrying with it men, churches, property and houses. The banks of the river were blocked and the water flowed backwards. This place was shut in by mountains on both sides, for the stream flows there through narrow defiles. The water then flooded the higher reaches and submerged and carried everything which was on its banks.

A second time the inhabitants were taken unawares, and as the accumulated water forced its way through again it drowned those who lived there, just as it had done higher up, destroying their houses, killing their cattle, and carrying away and overwhelming with its violent and unexpected inundation of everything which stood on its banks as far as the city of Geneva. It is told by many that the mass of water was so great that it went over the walls into the city mentioned. And there is no doubt of this tale because as we have said the Rhone flows in that region between mountains that hem it in closely, and being so closely shut in, it has no place to turn aside. It carried away the fragments of the mountain that had fallen and thus caused it to disappear wholly.'

Marius of Avenches also described the event in his Chronicle:

'The great mountain of Tauretunum, in the territory of the Valais, fell so suddenly that it covered a castle in its neighbourhood, and some villages with their inhabitants; it so agitated the lake for 60 miles in length and 20 in breadth that it overflowed both its banks; it destroyed very ancient villages, with men and cattle; it entombed several holy places, with the religious belonging to them. It swept away with fury the Bridge of Geneva, the mills and the men; and, flowing into the city of Geneva, caused the loss of several lives.'

According to one scientific opinion, it would have been a magnitude six tremor to cause such a landslip. And the evidence shows that this was not the first time part of a mountain had fallen into the Rhone valley, there were several smaller falls before this event, which, researchers showed caused a wave, probably higher than six metres to roar along towards the other end of the lake where Geneva is still situated today.

A tidal wave ravaged the banks of Lake Leman during the Bronze Age. Analysis of sediment on the bed of the lake reveals that around 1750 BCE, the people living on the banks of the lake also experienced an earth shaking event, leading to a sudden and violent eruption of a huge wall of water flowing along from the head of the lake. The result was a devastation of lake villages along the shores, subsequently abandoned. The cause of the wave was the backing up of the waters of the Rhone resulting from the slippage of the terrain and a massive rock fall from the towering mountains left over from the last glacial period, when the ice age gouged out the river valley. Techniques of geophysics were used for analysis, directed seismic energy, and the propagation of generated sound waves through the layers of the geological strata were used to explore deposits beneath the surface. Organic material provided a date range for the event, establishing that the material had been deposited nearly 4,000 years ago. It appears that an enormous mass of rocks and earth, over 100,000 cubic metres, was deposited 80 metres under the surface of the water, between Lausanne and Evian. This landslide was also part of a single mass of rock slide falling from the mountains down the steep valley walls.

Fast forward to the year 563, and remarkably, an even larger and more massive rerun of the Bronze Age event caused another catastrophe of such dimensions that it made it into the two European history books of the time. Discussion raged for many years over the exact location of Tauredun, however the two histories provide some clear indications that it had to be close to Saint Maurice, because the site had to be near the constricting passage to block the Rhone. Any further towards the lake, and the water could not have caused a tsunami, it would just have spilled out and absorbed into the twenty square kilometres of plainland at the head of the lake.

The Rhone was around ten metres lower than it is today, and because of the massive movement of the original terminal moraine upstream and downstream caused by the rockfall, the conclusion is that the Rhone flowed much closer to Les Dents du Midi in 563 than it does now. In depth geological analysis of the moraine, remnant vegetation, and the elevation of the river then and now, has produced the conclusion that in the first instance, the massive rockfall from Les Dents du Midi collapsed onto Tauredunum and blocked the course of the river. It displaced large quantities of moraine, which accumulated and formed a wall, blocking the passage of the water, creating an impromptu lake between the mountains. The water rose to a height of thirty metres, then broke through the moraine dam. A second slippage again blocked the river, and raised the water level another fifty metres. Once again, the water broke through and discharged a massive wall of accumulated water into the Chablais. The enormous force of the wave formed a tsunami, speeding over the Rhone plain, picking up alluvial material on its way. It continued to the lakeshore at Villeneuve, and then thundered all the way to Geneva. The geological evidence presented by Denis Fournier in the publication 'Le Tauredunum et la Geologie' dovetails very well with the account by Gregoire de Tours, which must therefore have been collected from eyewitnesses.

As water tends to do so powerfully, it washed away all the debris, the soil moraine, and rocky material that fell from the mountain, and distributed it throughout the Chablais and into Lake Leman. There are also huge blocks of rock lying around today at Saint Maurice that according to geological analysis, must have fallen from a great height during the Tauredunum event. It was the water casting the impermeable moraine against Les Dents du Midi that allowed the huge reservoir to form, and that prevented the water from just seeping away. In fact, because of the narrowness of the valley at this point, moraine obstructed two thirds of its cross section. Analysis of the rock has fixed the height of the dam rising up to 85 metres above the river bed before the collapse. The probability was that Tauredunum, when it collapsed, was perched above Saint Maurice, an extension of the plateau (where standing stones have been found) where Verossaz exists today.

The moment the waters of the lake found a way out of the moraine serving as a dam wall, tremendous devastation must have occurred. It is difficult to imagine a phenomenon of such a magnitude where an entire lake formed, until the water exercised its incredible power to break through. The water would have rushed into the breach, and its destructive powers were amplified as the wave picked up momentum, carrying rocky material piled up since the ice ages. All of the downstream region must have been ravaged by the wall of water. This was a cataclysmic event that caused a disaster of epic proportions, killing everything in its path, and disrupting the very fabric of life in the Chablais.

Evidence has been found that shows that the whole of Saint Maurice was covered in water that had backed up behind the blockage. During excavations for construction, a skeleton was found in the Rhone gravel at a depth commensurate with the event, covered by humus, then more gravel, then topsoil. A lone body found buried deep underground, from a seismic catastrophe, the remains of one single person so poignantly found after 1400 years.

Archaeological exploration uncovered the only known remains of any buildings from before 563. Located to the east of Saint Maurice, an underground vaulted arched room was excavated, probably a cellar, with walls 3.6 metres thick.

Mountains do fall.

THE LESSONS OF HISTORY

The Gallic Celtic people who lived in and around the Chateau de Tauredun, at the head of Lake Geneva, in what is now Canton Vaud, Switzerland were probably related, by dint of geography, to the La Tène people of Neuchatel 100 kilometres away, and perhaps also by ancestral culture, deeply connected to the people who painted the cave art at Lascaux around 19,500 years ago. As the Bronze Age gave way to the Iron Age in Switzerland, the Celtic tribes who circulated from Gaul east to the alps and westward to Britain, formed travel connections and relations, by language, and trade in metals and salt, ornaments, implements and artefacts. They formed ties of kinship, and bonds of commerce and friendship throughout the territories of the Celtic tribes through alliance with the people of ancient Gaul, and Helvetia, the Swiss plateau. These people lived on the alluvial plain of the Upper Rhone Valley, on the ancient trade route that connected the great mountain passes from northern Italy, facilitating the trade between the ancient civilizations with Gaul and Britain.

Like the rest of the Celtic world, they developed intricate geometric art based on precise measurements, and diffused and propagated their knowledge of astronomical events such as solstices and equinoxes, both as oral record, and in the mathematics of the design of artefacts. They left gold plaques, mirrors, urns, ornaments and other artworks in precious metals, with intricate engravings. They recorded solstices, and transport directions in the alignment of their public buildings, as well as in their trade and communication networks, making contacts and exchanging information, not only with Celtic tribes over vast tracts of western Europe, but also with the people south of the alps, the Greeks and Etruscans. The brilliant La Tène culture of the late Iron Age extended from the western Iberian Peninsula, throughout Gaul (modern France) and Helvetia (the Swiss plateau), west to Britain and Ireland, across southern Germany and east throughout Austria.

The alps of Helvetia, ancient name for Switzerland, hide their own secrets. To anyone who has stood and admired their magnificence, they clearly have some fascinating tales to tell, of lives and liberty, of tribulations and courage, as well as the sheer power and beauty of rock formations pushed up from deeper geological layers. The mountains originated from clouds of cosmic dust, generated in stars, and blown off by winds and stellar explosions, made of hard refractory minerals when the solar system was born around 4.5 billion years ago.

The ancestors weathered a very interesting series of historical events, when their civilised world was interrupted by a highly organised, opportunistic band of wealthy Romans, who, not content with the riches gathered from their own regions, conquered and colonised, and grew wealthier by force and enslavement of other tribal territories. Roman incursions into the Celtic Gallic tribal territory and finally conquest, was followed by the dark ages, the rise and fall of feudalism, and the forging of the republic of free men and women. How on Earth did these people survive? What motivated them to conserve their freedom by finding new places to inhabit, rather than subject themselves to tyranny? When and why did people begin to inhabit the wild valley populated by lynxes, wolves and bears, with freezing winter temperatures, off the beaten tracks that existed at the start of our modern era. And what lessons can we learn from our ancestors, to apply to the greatest moral challenge humanity has ever faced, climate change.

When the Savoyards finally colonised the Ormonts Valley around the 12th century, the family Tardent, French speaking people until they became polyglots in the 19th and 20th centuries, were present, found in early accounts of modern Swiss history. The French pronunciation is identical to Tauredun, or Tauredunum as it was known at the time of Gregoire de Tours, and Marius d'Avenches, monk historians of the sixth century. A town, once a fortified settlement with roots in the Iron Age, was completely destroyed by the collapse of the mountain above it, with such violence that it collided with the neighbouring escarpment, crushing the chalets, people, goods, chattels and animals under thousands of tons of rock. Those who were not buried were swept away by the flood.

The event not only took out the inhabitants of the Tauredun settlement, but also the villages above and below, on the Rhone plain, either drowned by the backing up of the Rhone as its course was blocked by massive slabs of rock, or by the huge wall of water that found its way through the blockage, roaring like a train to engulf the villages in its path, as the Rhone became a tsunami, a torrent that swept everything in its path on the banks of Lake Leman, all the way to Geneva over 50 kilometres away. This was an event of seismic proportions, both figuratively and literally, as the 60 days of rumbling by the mountain before its fall must have been related to a subterranean shifting of the rock mantle girding the Earth, to have caused such a dramatic collapse. An event that is irrefutable in early European history, it was recorded independently twice in the sixth century. Recent explorations of the bed of Lake Leman, the lake of Geneva, indicate that similar, though not as massive events occurred in the Bronze Age, roughly 3,800 years ago, according to dating of remnant vegetation.

The original site of Tauredunum has been determined by archaeologists and geologists, on what was a strategically sited hilly plateau above the fertile Rhone plain at the head of the lake of Geneva. One geological survey dovetails almost exactly with the account of Gregoire de Tours, and this means that despite the devastation, there were eyewitnesses on higher ground who observed the event. It is comforting to think that some inhabitants realised that a loud rumbling from a mountain over an extended period of time was a clear signal to decamp to safer ground, hopefully in time to take with them, their household belongings, animals, and eventually, the name. The family has been writing its history since the end of the 18th century, out of general interest, and as part of a habit of record keeping from earlier mountain chatelains since the 15th century, responsible for the establishment and keeping of communal records in the Ormonts Valley, that runs into the Upper Rhone Valley, not far from the head of the lake. An historical account by Henri Tardent provides some illuminating nineteenth century French Swiss political, historical and environmental insights.

So how does our newfound knowledge of the universe and the development of human cultures in different parts of the planet relate to the ancestral past of our Swiss family? They have their roots deeply embedded in the Gallic Celtic tribes of what is now Switzerland. The name, directly attributable to an Iron Age fortress above the valley carved out by a massive Rhône Alpes glacier, is a good indication that this family were not incomers, but part of the Celtic world linked by trade, marriage, language, culture and technology to the Gallic tribes. According to the limited number of ancestral mitochondrial haplogroups in Europe before the Roman invasion, the people of this region must have had close connections and links across the communication routes of the Celtic people, particularly to neighbouring Gaul, and the Helvetic plateau. What can be learned from their response to changing climate, political and economic conditions?

PERSPECTIVE

Earth is the spaceship in which we ride with our galaxy on an amazing journey through the universe, at a speed of over 500,000 mph, while the Milky Way, although apparently being pulled towards Andromeda, is actually dancing to the cohesion of a massive gravitational force balancing the expansion of matter. In reality our blue planet is located at the tip of a spiral arm of one of the massive neural networked superclusters, Laniakea. This much we know.

Dating the Earth is a fascinating topic. We have been able to piece together a timeline with a point of origin, an approximate estimate of the age of the solar system of 4.5 billion years, based on radiometric dating. The age of rock can be estimated because we know that radioactive elements decay at a certain rate. Radioactive isotopes produce concentrations of decayed elements, when compared with initial concentrations, because the known rate of decay (half-life) provides the inputs to calculate the age of rocks. Because the radioactive material is confined within the rock itself, by sampling enough rocks, a view can be taken on the age of the Earth, given that the initial parent element and all the decayed elements are all still present, trapped in the physical bounds of the rock structure. The inference is that the Earth formed as a planet at the same time as the rocks were formed.

From about two-and-a-half million, to twelve thousand years ago, there were repeated ice ages on Planet Earth. They were predominantly caused by orbital changes of the earth around the sun, affecting the total amount and the distribution of sunlight reaching us. (Today's climate change, in contrast, is largely being driven by humans increasing the level of gases providing an Earth blanket in the atmosphere, producing the greenhouse effect that initiates global warming). Because we are subjective creatures, with senses that have developed in response to our environment, we view the cosmos as starting with the formation of our planet and our Moon, both part of the net balance of expansion and the gravitational attraction amongst physical elements of matter over time.

We experience our universe as the resultant energy from a great spatial spin machine in which clumps of matter collide with others, in a vast rotational dance of mass. Matter is a projection of pure energy, the structure of which has behavioural quantification governed by mathematical principles. Since the surface of last scattering of energy plasma, at the beginning of the universe as we know it, over vast aeons of time, matter consolidated from predominantly hydrogen, forming helium, the lighter elements and subsequently the heavier metallic elements, which bonded together in a mad cosmic dance to form congregations of space debris. Clouds of interstellar dust congregate to form stars, transformed by immense energy flows and explosions of luminous radiation.

Eventually, in our galactic supercluster, matter stabilised as large irregular shaped meteorites, comprising carbonaceous material, and water (as ice). Our solar system was formed by the ineffable universal forces into roughly spherical bodies, by dint of friction and rotational spin, shaping the net gravitational mineralized rocky material into our planets orbiting our star, the Sun. We arrived at a point of entropic balance, a most improbable arrangement of rocks, gases and ice, from which by chance and mathematical design, amino acids formed the first cellular protein machines from which all life as we know it, gradually evolved on our blue planet, Planet Earth. The greatest mystery of all time.

Most stars are in a plasma state, as is our sun. Plasma is the fourth state of matter, a prelude to gases, liquids and solids. Plasma is ionised atomic matter, containing charged particles, and has a low probability state of entropy (disorder). That is, it has a high energy state. In the vast empty parts of space between galaxy clusters, there is still a rarefied form of plasma. No part of our known universe is a complete vacuum. The sun, like every other star, formed as a result of the initial expansion of the universe. Cosmic microwave background radiation (CMB) is a remnant of the initial energy emanation. Once light, it is now in the infrared range because as space expanded the wavelength increased. It still permeates space and comprises around 1% of white noise. Theories of the early universe now include a variable speed of light to account for both the shape of the universe and the uniformity of temperature and density of matter. Initial quantum energy fluctuations are thought to be the patterns responsible for the distribution of the elements that today we see as stars and galaxies.

Light travels in straight lines. Or does it? There are no straight lines in nature. The natural world that we humans experience, runs on energy that emanates from an entire universe of neural networked galaxies into the dissipative edge of the supercluster Laniakea, where the Milky Way resides in a fragile balance of turbulence at the edge of chaos, not far from one of the vast empty spaces in the cosmos, interspersed between galaxy superclusters, where there is practically nothing. Space is expanding, and counteracting the expansion is gravity, the result of matter interacting with matter at a precise time, the present moment. The Milky Way, like other galaxies, is being pulled towards a region of our supercluster known as 'The Great Attractor' by astronomers, (recently dubbed Laneiakea, Hawaiian for 'immeasurable heaven'). There are billions of stars in our galaxy, of which our sun is one. And we, on our beautiful blue planet, are orbiting the sun approximately once a year, as our spiral galaxy, one of billions, moves rapidly through space being pulled towards the great attractor at the virtual epicentre of our galactic supercluster.

A beautiful piece of rock, pitted and cratered by collision with asteroids, planets and other moons, by coincidence presents always the same face to us here on Earth. It is our one and only moon, being pulled along for the ride while it orbits gracefully around us once a month, with some phased beams of light refracted from the sun showing us the way to travel at night. The moon has its seasons as well as the sun. When the Earth is at a particular inclination, so that moonlight is partially or fully obscured, it is the season of the lunar eclipses. The moon, for us, is like the second hand of a watch, and as the sun tells the minutes, the milky way the hours, it is the galaxy clusters and superclusters that tell the weeks, months and years.

Newton discovered a set of laws which work within the context of our solar system, for the attraction that one piece of matter holds for another. Einstein discovered that the underlying basic premise of light travelling in straight lines, does not hold true all the time. The anomalous perihelion of Mercury indicates that something else is going on, other than gravity working as though it were a mysterious force at the centre of the planets and the sun. What is true for the solar system, does not stack up for the galaxy. Light is bent by the net energy resulting from the interaction of matter. And so Einstein's theory of general relativity was developed, premising that the spatial geometry of matter and time is warped energetically. Theories of dark matter and black holes are being challenged by cosmologists and physicists who believe that we do not yet have a theory of gravity that works for the universe.

Hypotheses and new theories are being developed that link quantum mechanics, with a universal spin force, and a new view that the speed of light may not always have been a constant. Physicists have not yet found a reason for the arrow of time, so the question is still open, it may or may not be reversed. Perhaps it is equivalent to constant multi-dimensional momentum, in harmony with the spin that causes the spirals of matter evident in the cosmos. Time is not yet fully measured or understood, clearly it is not simply a linear interval between two events.

So many poets and musicians have issued paeans, hymns of praise, to moonlight. So many belief systems have put the sun at the centre of all things, and so many believers have worshipped the sun since the dawn of prehistory. The rhythm, phrase, colour, tone and beauty of the universe are enchanting. Either we need to get out more, or we need to fix our ecosystems so that it is possible for everyone to see the stars at night.

There are no straight lines, however the measurement systems we humans have developed are wonderful inventions of the human brain. We have devised many clever systems with logic, rationality, and numbers. Analytic concepts are our real tools, which we've developed so successfully over the millennia of proto-history, and the two thousand odd years of written anecdotal evidence of recent times. The ability to envision technology, and the persistence to follow through in the realisation of the materials and development of new ways of improving our collective living conditions is singularly the greatest, real achievement of humanity.

We have abstracted ourselves from our environment in an alarming way. We do not follow the sun, are unaware of the phases of the moon, most of us cannot synchronise our eating habits, let alone our horticulture with the seasons, and have no idea of what we and our fellow inhabitants of the third rock from the sun are doing to the planet's biosphere. The oceans continue to warm and acidify, and global mean sea level to rise, and even if we stopped greenhouse gas pollution immediately, we cannot now stop the ice sheets and the glaciers melting at continuously faster rates. There is a great deal of uncertainty as to what is going to happen to our global weather systems, but one thing is certain, the current resistance to addressing climate changes due to burning fossil fuels in the atmosphere by vested interests has to change, or we are in for a very rocky ride.

It seems that the balance of the universal forces that shaped our planet is self-regulating, according to the Gaia theory of James Lovelock and others. The atmospheric balance that regulates the production of life on Earth currently provides a gentle cradle for the development of human civilization. How amazing! And how sad that we are meddling with the stasis of the air we breathe, the rain that falls to nourish the plants, the forests that provide our oxygen, and the acidity of the oceans that are the prime initiators of recycling greenhouse gases.

Chaos theory, life's governance mechanism, provides the laws of all systems composed of the building block elements generated in the stars, i.e. carbon, hydrogen, oxygen and nitrogen. This theory teaches us to expect the unexpected! A small change in the initial state of a system provides dramatically different physical outcomes. When there is a positive feedback loop present, the change from known, predictable limits is accelerated. In this context, the term 'chaos' does not mean confusion and disorder. It refers to the loss of stability, as systems oscillate wildly from one state to another. Chaotic behaviour can be observed in many natural systems, such as weather and climate. An example of a feedback loop is the melting of the polar ice caps. Because there is less white surface to reflect the sun's rays back into space, more heat reaches the ice, and it melts faster. We are on the beach of climate change, watching the waters slowly rise as the poles slowly melt. We owe the ancestors this much at least, to discover as much as we can about their collective journey, so that we can unravel any past mistaken directions, and regather the threads of reason and enlightenment to weave the next chapter of our human existence.

EARTH DIARY

Perhaps the big mistake we make is always to presume that the people of history were lesser beings than ourselves, somehow more primitive, more superstitious, more ignorant and of lesser intelligence. The current predicament of dwindling resources, looming climate cataclysms, overpopulation, and our attempts to problem solve using conflict show this premise to be outright self-deception. In times of past crises, when humans have been nearly wiped out as a species, how did our ancestors manage to survive? After all, analysis of mitochondrial DNA and Y chromosome sample populations show that we are all descended from one common male ancestor and one common female ancestor, who lived at different times somewhere between 100,000 and 150, 000 years ago. At this period in our evolutionary history, the human species must have come perilously close to extinction several times through natural events and disasters, for the respective progeny of two single individuals to provide our common ancestry, before populations picked up again.

Modern humans have existed approximately for the last 200,000 years, and a small percentage of our

DNA comes from several other hominid haplogroups, in particular the Denisovans and Neanderthals. Over the past 150,000 years, humans developed a larger frontal cortex, providing the short term memory and processing that enabled us to become adept with a range of tools and technologies. Fast forward to the end of the last glaciation around 12,000 years ago, and groups of modern humans had managed in some regions to survive the extreme environmental changes, through adaptive strategies. It must have been an amazing feat to meet the challenges of the climate. Somehow, people had managed a lifestyle constrained by the rigours of a much colder planet. Being clever with technology would have been the key to enabling our ancestors to conduct their daily lives.

According to analysis of DNA performed under the auspices of the National Geographic genographic project, humans first ventured out of Africa over 60,000 years ago, leaving genetic footprints of their migration routes, still visible today. By mapping the appearance and frequency of genetic markers in modern peoples, the genome project has created a picture of when and where ancient humans moved around the world. Two major streams of these early humans were the ancestors of modern Europeans. The exit route out of Africa was through modern day Egypt, Iraq and Iran. Somewhere around Iraq one group of people moved north east, eventually branching to central east and south-east Asia, and the Indian subcontinent.

During the last Last Glacial Maximum on Planet Earth, human beings were slowly advancing throughout the globe. One migration route doubled back to the north west, across modern Poland and Russia, eventually reaching France, Spain and Britain. The other migration path to Europe was westward through the middle east, and via the Mediterranean, onward to modern day eastern Europe, Italy and Switzerland. Still another group ventured along the southern shores of the Mediterranean to north-west Africa. Yet another more southerly migration route out of Africa was via modern Ethiopia, hugging the coastlines of Asia and Indonesia, with the more intrepid eventually reaching Australia.

Since our planet began to support life three or four billion years ago, there have been at least five grand ice ages, periods of long-term reduction in the temperature of Earth's surface and atmosphere, resulting in the presence or expansion of continental and polar ice sheets and alpine glaciers. Between the ice ages, are the interglacial times, when the planet warms up again, and permanent ice sheets retreat to the confines of the polar regions, where the lower winter temperatures currently prevent a complete melting of large volumes of ice. The current ice age started about two and a half million years ago, when ice sheets once again spread across the Northern Hemisphere, where planetary ice sheet growth begins (because of the tilt of the Earth on its axis, there is, in total, less sunlight reaching this part of the planet).

Cycles of glaciation with ice sheets advancing and retreating have occurred periodically, lasting somewhere between around 40,000 and 100,000 years in duration. We have termed these cycles 'glacial advances', and 'interglacial periods' when the ice retreats. We are currently experiencing the warmth of an interglacial period, after the last glaciation from around 70,000 to 10,000 years ago. Interestingly, and unsurprisingly, this coincides with one of the grand scale migrations of humans out of Africa. Fossil records indicate that this cold snap would have encouraged our African ancestors to search further and wider for food and shelter. Around this time the number of human beings experienced a sharp drop, declining to fewer than 10,000 people, according to genetic evidence, and the effect of sudden loss of habitat for our ancestors would have meant that the few who survived did so by providence in finding shelter, and ability to use technology innovation with fire and stone tools to support food production. At least from that time onward, humans were using their talent for survival to migrate in search of opportunity for a better life.

Not everywhere on the planet was covered with ice like vast swathes of northern Europe (and of course Switzerland because of its alpine elevation) during the Last Glacial Maximum, the peak of the ice age. This was the period in the Earth's climate history when ice sheets were at their most recent maximum extension, between 26,500 and 19,000 years ago. At this time, the western end of the steppe-tundra zone extended from south-western France through northern Germany and the central European plain, across the Ukraine to the Eurasian steppe as far as the Ural Mountains.

During this period, the level of solar radiation (sunlight) and the quality of the silt sediment soil, provided rich ecosystems for growth of ground cover vegetation such as mosses, lichens, grasses, and low shrubs. This terrain sustained antelopes, mammoths, horses, and reindeer amongst other grazing animals. Our hunter gatherer forebears followed the herds, and colonised the environment, becoming expert at hunting reindeer especially. However, at these latitudes, though today they are temperate regions, they had to deal with winter temperatures as low as -30° C which must have been very difficult in those places where fuel and shelter were scarce. Our ability to innovate with technology is what allowed us to continue to survive over the millennia of the last great ice sheets. Our ancestors travelled on foot and relied on following and hunting the herds for food. Challenges of extreme cold were overcome through developments such as production of clothing from the pelts of animal fur, construction of shelters with hearths using bones as fuel and digging of "ice cellars" into the permafrost for storing meat and bones.

We clearly have an amazing ability to adapt to changing circumstances. As the ice melted, and our talent for creative use of natural materials led to the discovery of metal working techniques, we entered an age of abundance, where milder, gentler climatic conditions allowed people to access many and diverse food sources, with a variety of temperate and tropical ecosystems covering the globe. Today, all that remains of the continental ice sheets are the Greenland and Antarctic ice sheets, the Arctic region and the glaciers dotted around polar and alpine regions. It is only 10,000 years since our planet became much more temperate and tropical, capable of supporting the teeming and biodiverse life forms that nourished human beings. We learned to farm and preserve the fruits, vegetables, fish, wildfowl and game that nature so bountifully provisioned.

And yet somewhere in this Earthly paradise, were seeds sown that by the late Iron Age spawned the madness of war, conquest, enslavement, and the violent appropriation of food and water by force from other tribes of people. As a species, we have been slow to learn to consciously manage the natural resources that support us. We are still in the very early stages of learning to share and collaborate about goods and services on a sustainable economic basis, placing a value on natural resources, fair trade, equality and mutual advantage in all forms of partnerships, including marriage. While certain technologies enabled an easier living, when resources become constrained, we failed to heed the wisdom of the ancients, to invent, use and produce technology to solve local problems. Instead, somewhere in prehistory, we started to use our new-found abilities, skills, and knowledge to aggressively and violently appropriate other people's resources. Correspondingly, we have a poor record as a species, for sustainably managing the natural resources that provide for our very existence.

When resources become constrained, as they inevitably do with the pressures of increasing population, we begin to take our neighbours land, goods, services, and technologies, acquiring them by use of force. The temptation to use weaponry and war as an occupation of choice from the vantage position of horse and chariot proved too much for the people of the Eurasian steppes. This was not the only fault. While the advent of farming had provided ready food and the ability to remain in one place in an organised settlement, the civilizations of Mesopotamia flourishing around the Tigris-Euphrates river systems eventually failed. It was not just territorial incursions from neighbouring tribes. Natural resources, soil quality, water, and overuse of terrain along with unchecked rises in local populations all led to famine and hardship. Eventually the earlier civilizations were overcome by incursions from the Scythians from the Eurasian steppes who organised a system of conquest, rewarding warriors with the spoils. By the time of the late Iron Age, conquest allowed them to establish a stronghold in what is today the Crimea. From the glorious rise of towns and civic organisation, it was only a matter of time before the Iranian Plateau became a theatre of war, which has been followed by the rise, decline and fall of a succession of dynasties ever since. After the Scythians eventually fell, they were followed by, amongst other regimes, the empires of Persians, Mediaeval Islamists, Arabians, Mongols, Turks, Shias, and in the 18th and 19th centuries, European colonists. Ongoing political unrest plagues the region today, and environmental degradation and uncertainty of people's liberty is clearly the price of war.

Perhaps there are some lessons to be learned from the succession of warlords in this region. Seemingly once the genie of organised invasion and conquest was released from the lamp of the Iron Age, it has proved very difficult to put it back. Perhaps there are some lessons to be learned about ourselves by studying the animals that are able to live in balance and harmony with the natural environment, without the compulsion that seems to bedevil our species, wanton destruction of our own species and our natural resources in the name of progress. At this moment, the rhetorical question has to be asked, are we in danger of throwing away our collective future, because of an obsession with money, power and glory, so little of it earned? In fact, are we spending the planet's resources at a rate that is faster than the planet's resilience? Are the mass disappearances of species and populations of flora and fauna inevitable, or can we save the Earth from the extinction on a massive scale, of plant and animal species on the planet, including ourselves?

In prehistory, long before the Roman empire, archaeological and historical evidence indicates that people and households respected a pantheon of gods and goddesses, providing for a wide range of cultural protection. The same evidence shows the large-scale development of weaponry against fellow human beings is quite a recent phenomenon, made possible by the production of first bronze, then iron. Earlier agricultural and hunting implements, tools, art and ornaments, and sophisticated methods of metalworking began to be applied to weaponry. Serious production of lethal swords, knives and shields appear to begin from around 1300 BC. Subsequently people discovered the process of iron smelting from iron ore, and the techniques remained basically the same till the middle ages. Iron oxide bearing rocks are roasted and crushed, and placed in a furnace to burn off the oxygen and produce blooms of wrought iron. From there it can be worked into tools.

What circumstances made people develop weapons on a large scale as well as tools? As the Bronze Age gave way to the Iron Age, the escalation of the arms race took hold on our collective psyche with a vice-like grip that today not only holds us in thrall, but threatens our very survival. It is clearly our own fault. In fact, it is only fair to comment that the escalation of warfare amongst humans seems to have generated an increase in superstitions and improbable belief systems. We accorded the owners of metal mines and technological processes an extraordinary status and wealth, and instead of everyone becoming versed in the knowledge and technologies of metal-working, there was a specialisation and a decrease in public interest in science and technology. The rest was a history of mining and metals trade that today is the cornerstone of a global labour and capital system. Was there ever a co-operative approach to technology and trade? Maybe not. While the Celts managed to develop their civilization largely through trade, marriage, language and the development of new technologies of transport and navigation, they also fell prey to the conquest of other people's territory, and by 250 BC it appears the tribal alliances were beginning to disintegrate as competition for resources, land and metals overrode common social practices and courtesies. It appears that the Celtic Hallstatt and La Tene cultures were equally obsessed with weaponry.

Sadly, it appears that intertribal skirmishes weakened the earlier bonds. The Celts, in the end, were perhaps their own worst enemies, failing to understand the nature of their predicament. Even though they used the old alliances to unite against the upstarts, they probably also failed to realise the extent of the threat, nor recognize in time the ruthlessness of the Romans. The lack of strategic cohesion across Celtic Europe meant that increasingly power obsessed and ambitious members of Roman society could divide and conquer. Fortune favours the bold, and the Roman army was successful largely by a combination of good luck and the overarching self-belief of their leaders, as much as anything else. The Roman rulers clearly spent some time plotting and scheming tactics and strategy that would eventually lead them to raise an army of brutal conquest, the wholesale killing of men, women and children, and the enslavement of a large part of the European Celtic and Germanic world. Caesar's megalomaniac exploits remained unrivalled in savagery until Hitler, and as emperor, he laid wholesale claims to the technologies of the people conquered. History was written by the victors, and to this day, populist European history appears to propagate a collective belief that attributes all the technology and cultural achievements of the first century BCE to the Romans. This is a complete misinterpretation of historical events, most probably propagated by Roman historians, many of whom were in the pay of the state, mindful of the need to please their political factions and emperors. Roman historians modelled their work on Greek historians, the basic form of which was writing an annal, giving a subjective account of the year's events. To give the Roman historians their due, in the true tradition of writers, many used veiled allusion to hint at truths that they were perhaps constrained from telling due to political influence.

And yet in the twenty-first century, all of our irrational responses to the crises and difficulties we face seem to stem mostly from our inability to analyse our own position. Our ineptitude to deal with global crises and conflict stems from our lack of understanding of our dependency on our natural environment. Fear generated politics, and our famous ability to follow any leader who promises us the Earth have often been our downfall. We have been looking for a quick fix, a ubiquitous hero to fix our lives, and lead us to glory, instead of understanding that our difficult situations arise because of the pressures of declining availability of land for herding and agriculture, the scarcity of food and water attributable to population growth, and our seeming inability to apply our brilliant technology gifts to solving our deepest problem, war.

An earlier round of global climate change is reflected

in climate models that seek to understand past global ecosystem instabilities. The desertification of northern Africa is theorised to have resulted from changes in Earth's orbit, causing feedback loops to amplify regional changes in microclimate, vegetation, and water sources. Changes in territorial behaviours accompanied the climate change that caused the desertification of the Sahara around 6,000 years ago. This was the trigger that caused the migration of sub-Saharan Africans to settle along the fertile banks of the Nile. So what has happened to us, that we have developed such a predilection for hero worship for the conquerors? Why does the critical mass of people still look to a political patriarchy for promises of victory, rather than a rational assessment of the facts? Is this a result of a loss of self-confidence because of a deep psychic fear resulting from past wars?

The Earth's human cultures have not always been patriarchal and authoritarian by nature. Africa, Indonesia, Asia and New Guinea still boast matriarchies of ancient order. Are our cultures so obsessed with authority and process, physical power and control of others, rather than respect for the group and the value of every person's contribution? Why are we so good at taking individual responsibility and problem solving for mutual benefit, and yet as a society we are acting under the hypnotism of global monetary and financial systems? Why do we undervalue our right-brain emotional nature, which governs our softer responses? When did our collective thinking get so out of balance that we developed an unhealthy preoccupation with conquest, and the killing and enslavement of other human beings for the fleeting rewards of unearned status and privilege? Why do we not object more strongly when we see others profiting from people's misery, when we know that fear based decision making is never going to produce positive results? What have we got to lose, except the continuity of human civilization, in the face of the world's sixth mass extinction event? These are interesting questions that deserve some consideration.

How can we overcome past conditioned responses, to answer the one question that is on so many people's minds today? Can we manage a concerted global response to the effects of playing with the Earth's climate, when we know without doubt, that the release of excess greenhouse gases is pushing the Earth's atmosphere, oceans and land masses to a tipping point, possibly of no return, in the foreseeable future of the human habitat? We do live in a world where the four corners of the globe are affected by climate system changes to rainfall and drought somewhere else. Just as for financial systems, if one country experiences extreme weather, the consequences are becoming increasingly unpredictable, and we are unable to forecast the scale and extent of the effects in other parts of the world. Humanity has to act in a global collective sense, cooperatively and collaboratively to devise tactics and strategy to address inhumanity, the destruction of the natural environment, and to apply our collective bent for innovation to curbing dangerous greenhouse gas emissions causing extreme climate events. We have to stop hiding behind the cloaks of nationalism, and enlightened self-interest, other names for greed. We have to persuade our wealthiest individuals and organisations to let go of the illusion that taking other people's resources leads to a winning outcome.

Sadly, the emergence of large scale weapons production and militarism seems to have a parallel history to the populist decline of the status and role of women in society somewhere around three thousand years ago, when we developed an obsession with the reliance on force to achieve objectives. Of course it is a paradigm that has served no-one, leading, as it has, to even more manipulative power plays by both men and women. Today we experience unprecedented levels of ruthless and unprincipled pursuits of money for the power it wields, the exploitation of psychological weakness in the labour force, and downright enslavement, in every country in the world, including the wealthiest nations.

Conflict and dispossession of people from their traditional lands has always led to an upsurge of a desperate search for meaning. The increase in stress caused by homelessness and statelessness usually results in the suspension and interruption in the development of technology and knowledge based on traditional wisdom. Eventually the mid Iron Age dispossession, and forced exile from traditional lands because of competition for scarce resources was accompanied by a rise in monotheistic belief systems, able to be traced first to Judaism in the sixth century BCE, and subsequently embraced by both Christian and Islamic traditions. It is no coincidence that the emergence of monotheism and the predominance of patriarchy took place around the populous fertile regions of the ancient civilizations.

Developments in farming methods and increasing populations eventually led to an associated decline of land and resources per capita. Combined with invasion and conquest of those who envied their progress, decline and fall of these cultures was inevitable. Part of the problem was due to the fact that there were no large natural geographic defences against incoming raiders in these regions. By contrast, Egyptian society had been surrounded by desert since its inception, with limited vulnerabilities to invaders. Religious belief remained polytheistic, with deities whose behaviour was often very human-like. A powerful priestly class helped to retain the same basic forms of worship for millennia throughout antiquity. One clear advantage for stability in Egypt was that rainfall, and the flooding events of the river Nile over this period were usually largely predictable, meaning that food sources could be anticipated and managed in advance. In contrast, competition for arable land, and the development of more sophisticated weaponry increased the invasion of geographically vulnerable established settlements by raiders and pirates. However even in ancient times, trade and transport meant that the local economies were inextricably linked from one state to the next. And eventually Egypt too fell upon hard times, from both the weather and invasion.

Yet, even after another three thousand years of bitter

experience, and a planet on the brink of environmental catastrophe through human induced climate change, we have yet to find a way to halt militarism and the taking of other people's lands and resources in favour of a balanced, healthy, productive society and environment. Our focus has often strayed from the careful management of the available natural materials and the ability to maintain energy and water resources. The best use of local environments is still undervalued, not properly understood, and worse, is missing in action from most of the world's education systems. The regrettable decline in the status of women has to be reversed, with women reinstated to their rightful place on this planet, as equal human beings in every possible way, spiritually, emotionally and physically, as we cannot afford to waste half of our intelligent contributions to technological advances. In short, we have to ensure that as a society we acquire new positive attitudes, not only with respect to people, but also to plants, animals and the natural environment. Everyone and every living being deserves to achieve their full potential.

What has humanity gained from the culture of warfare and conquest over the past 3,500 years (since archaeological finds have included large scale weaponry to be used against human beings)? The question really is what humanity has lost, during the periods of warfare, and the dark age that post-dated the fall of the Roman empire. People who lose their cultural identity always suffer from disconnection, confusion and the associated arrested development that accompanies loss of self-determination. This is a pattern that we have repeated through modern history, recent colonialism, and sadly it is still thriving in the present day. Writers, poets and artists have continually pointed out this dichotomy throughout the ages, with brief sparks of illumination for the black night of the soul engendered by loss of culture and traditional values. Under the guise of realism, in every generation there always seems to be a new group of people willing to forget moral values for the lure of wealth and control over others.

Declining land availability, traditional food sources affected by periods of climate instability, as well as competition for the increasingly scarce resources of water and agricultural land - together these factors resulted in behaviours where people took territory by force. Fear and bitter experience of dispossession of clan groups from traditional lands seems to have led to an enduring psychological cult of the warrior hero, both consciously and subconsciously, bound up with ideals of masculinity and femininity that serve the propagation of the myths of the heroism and glory of war. Sadly, the pattern of dispossession of traditional territory, and the escalation of weaponry (a proto arms race) repeated itself during the Bronze and Iron Ages, and indeed throughout modern history, ad infinitum, ad nauseum, and still exists today.

On the other hand, we also have to give ourselves a great deal of credit for our successes. The widespread emergence of the Bronze Age, where the smelting of copper and tin produced a range of tools, (as well as weapons), is evidence of the human propensity for adopting and adapting new technology. And if any indication is required as to our aptitude for innovation, it is provided by the transformation of the Neolithic population drift into long-distance travel and inter-regional metal trade routes, creating a new globalised economy connecting regional and local co-operative ventures. The organisation of labour and resources and the emergence of towns was no doubt a complex mix of individual effort, group dynamics, and discovery of new technologies, as well as the emergence of a new set of power structures amongst people. These social changes were intimately connected with the development of metal artefacts and mining, smelting and manufacturing processes as well as trade in foods, wine and other fermented beverages, and of course salt for preservation of food products. We just have to learn to put a value on the natural resources that the planet provides so abundantly, and to prevent war at all cost.

The pattern of technological advance, population growth, territorial incursion, and depletion of natural resources by farming in permanent settlements, repeated itself throughout the early societies and tribal groups across the geopolitical boundaries of both Europe and Asia. It has provided humanity with mixed blessings. There are of course clear advantages of cohabitation and the pooling of services to enhance lifestyle. However, in every civilization that has arisen since the Bronze Age, this has been accompanied by the disadvantages of single points of failure that accompany over-specialization. The biggest loss is the neglect of traditional knowledge of maintaining the natural environment. Living in towns and villages has always loosened our connection with the local ecosystems, the animals and plants, a connection which has been the driving force for our evolution as a species. Settlement has always resulted in a series of decline and fall of civilizations when our environmental resources became too unbalanced. Yet throughout all the upheaval, our societies have somehow clung on to the thread of gold woven into the fabric of our collective existence, the values of decency, commonality, and progress for all, in the face of wholesale cheat and deceit by a few people focused solely on the acquisition of power and pecuniary interests.

Accompanying our evolution over the past few thousand years (a drop in the ocean of evolutionary timescales) and the mastery of metals, is a darker flip side. The commercialization of exchange and trade based on mineral wealth, money as a currency for urbanisation and the adoption of unevenly priced capital and labour, has always resulted in wealth and power being concentrated in too few hands. Clearly there is something wrong with the governance mechanisms, and this paradigm has to change. It is a pattern that has been established for so many millennia that people have come to regard it as a normal state of affairs. Inequality does not further an evolutionary path for humanity, a fact we ignore at our peril. We are a species, a sum of our individual skills, working together collaboratively, and our sustainable evolution depends on raising standards for all people, not just because it is the decent thing to do, rather because we are all interdependent, we are all part of a global consciousness that requires all its parts, every single member of the human race, to progress and function as intended. We are a part of living, evolutionary biology that has to grow and transform with the times to fulfil its purpose. Like it or not, we grow or we go.

Written and oral European histories, at least in those early societies that evolved to occupy agrarian settlements, indicate that human behaviours under stress have declined into conflict and other unhelpful patterns of interaction, such as slavery and the concentration of resources in the hands of the few, the unequal division of labour and distribution of rewards. This is something not yet viewed totally objectively, as politicians continue to sway people with persuasive arguments about why, disguised as patriotism and nationalism, war and conquest is a good thing. In fact, the acquisition of money for its own sake, the insatiable desire for profits by the shareholders of super corporations does not benefit society. On the contrary, it also propagates a psycho-emotional entrapment to the endless pursuit of money. Money is a means to an end, and it has to be used wisely because it is a form of energy. Hoarded and objectified as an end in itself, it merely blocks the natural flow of goods and services. Obsession with money inhibits the exchange of skills, knowledge, resources and goods amongst people, in accord with principles of fair trade and natural justice.

Money is, in fact, a very useful current that provides

for the realisation of power and the potential for energy usage in all its forms. It has to be measured and managed to meet the supply and demand of natural resources for us all to use and enjoy. A clear, ubiquitous value has to be put on the Earth's finite water supply and all the minerals, plants and animals, the natural resources of the environment. In addition, a fair trade for paid and unpaid labour is essential, as is the sustainable use of 'reinvented fire', energy for manufacture, electricity and transport. And we have to move away from a world where cheating and deceit are encouraged as a normal way of conducting business. As a past (and present) response to the emerging disparity in standards of living, some people have developed an obsession with status quo, not based on achievement in science, the arts and humanities, technology and culture, rather on wealth as a means to display power. In most societies, there is some form of virtual entrapment of labour as a practice for accomplishing work, not for the common good, but for private gain. Lack of equality and disparity in standards of living is not, and never has been beneficial for a society. It is divisive, creates social unrest, and reduces spending on social harmony and the development of innovative technologies.

The current trend by world governments to encourage the use of physical force for the purposes of controlling fellow human beings is an abhorrent practice, something that creates anti-social behaviour that is not only ethically undesirable, but also causes enormous economic and social costs that far outweigh a short term advantage gained by a few people. It is cheap profits at the expense of the Earth's resources, and the commonwealth of us all. Our magnificent intelligence, highly evolved and specialised, deserves to be utilised creatively to its highest capacity. Rather than paying us to be cogs in a wheel in an endless rinse cycle, where the deficit in human potential grows day by day, our economy has to change so that innovation and hard work are rewarded, and a real value placed on the planet's ecosystems, fostering a sustainable development and renewable energy based global capital economy.

By contrast, the development of wealth from trade when shared fairly, provides an impetus for all people to live in balance with the Earth's biosphere that we share with animals and plants. Equality has always been accompanied by a rise in the common standard of living. On the other hand, since the Bronze Age, the concentration of wealth in a few families, achieved by over-regulation of human behaviours has always occurred immediately prior to the decline and fall of civilizations. So many times cultural knowledge and art-forms have been lost, in different parts of the world. So many of our technological developments, works of art, scientific methods and cultural lores have disappeared through a collective obsession with power and money, to the exclusion of healthy interaction with the natural environment. Release of our apparent entrapment by, and addiction to the forceful use of power, can facilitate a state of harmony and balance with natural resources, needed to facilitate the evolutionary leap for humanity to achieve a more prosperous, sustainable, egalitarian future.

We can change our education systems to produce free thinking, innovative, happy, confident human beings who respond eagerly to the challenge of problem solving for a prosperous equilibrium where we live at peace with ourselves, with one another, and the planet's ecosystems. The alternative is unthinkable, as our inability to respond to climate change in a timely coordinated manner has already shown. We can awaken from our current state of environmental apathy to curtail what is already proving a period of extreme weather instability. We must act to prevent wide reaching social unrest and knee-jerk authoritarianism in the face of inevitable changes to Earth's weather systems. We have to improve disaster management, our response to the risks inherent in climate change, and make use of our ability to forecast loss and damage from polar ice melt. We have to pool resources and pull together to prevent the large-scale displacement of populations by rising sea levels, adverse weather events and territorial aggression, so that we can avoid another long dark age, this time with the Earth's natural resources depleted to the point of exhaustion in many parts of the world. Really, unless we take concerted action urgently, there may be no comeback this time round.

The wholesale destruction of the forests, the lungs of the world in South America and South East Asia, the desecration of local land with toxic residues from mining, fracking, transport, and shipping are clear evidence of something awry with populist culture and belief systems, which is accurately reflected in our governments. The world's oceans are being fished out, including slaughter of our mammalian cousins, dolphins and whales, by large industrial conglomerates, forcing out traditional fishing, as well as islands of plastic such as the Great Pacific Garbage Patch, where ocean currents form a gyre that has concentrated plastic debris just under the surface. Similar areas are to be found in the Atlantic. Fossil fuels are generating more greenhouse gases than the atmosphere and oceans can absorb. The continuing devastation in Iraq of culture, values, people and lifestyle of the past couple of decades of intervention under the cloak of preemptive warfare, is surely testimony to the observation 'insanity: doing the same thing over and over again and expecting different results'.

So as we begin to recognize that we are living on borrowed environmental time, can we pick up the threads of a proto-history such as the Celts once had, when people expanded territory and knowledge via trade, language and marriage? Can we engage the attitudes and behaviours that we displayed prior to the infighting and wars that led to the conquest and enslavement of the Celts by the Romans? Can we recognize the earlier behaviours that led to a process of cultural decline through infighting and rivalries that led to internecine destruction? This is a pattern that has been repeated over the past two millennia and more, as all empires, including the Roman, the European and British colonialism, and the Soviet Union, have eventually declined and fallen. We are witnessing the decline of the United States cultural imperialism, and hopefully China's regional territorial invasion and extreme authoritarianism is to be short-lived. Australia's environmental destruction of the past two centuries by misapplication of land management to a fragile continent is hopefully slowly being halted. We do not need unfettered exploitation of the Arctic as a response to reduced sea ice. We do not need mining and fishing on a large scale in Antarctica. If we can recognize our own collective stupidity, and change our automatic responses to the dominant paradigms of populist culture, can we begin to manage natural resources sustainably and successfully for the common good? Can we reject unwarranted aggression for the benefits of a kinder, more compassionate society?

Are we able to recognize our own historical patterns of denial? Or do we just continue to believe whatever we read about ourselves? Can we question our human behaviours, or do we remain ignorant of our own psychology? What does it tell us, that we have propagated the Roman view of history for two thousand years, without question? Perhaps we might reflect on who we are, and for what we really stand. Are we leaders and protectors of the planet, or are we ego-centric destroyers of life? That is essentially our stark choice at this moment in the history of humanity on Planet Earth. The truth is that we have lost our immediate connection to nature, with the potential for serious consequences because the planet's population growth and overuse of natural resources is today so much greater than at any other time in history. In fact, in our current era, the most fragile and vulnerable cultures and countries are still being colonised and unjustly exploited for mineral wealth, and land. We are exhausting the wealth of the natural environment at a faster rate than the Earth can replenish it. We are at a point of ecological instability that may prove increasingly untenable for our very survival. The roots of our current dilemma lie in the transition of population migration in search of opportunity to agrarian settlements, mining, and commercial trade, and the associated rise of the super-rich, and the decline of resources for, and exploitation of people existing in the location. This transition has been millennia in the making, and we have not yet dealt with the associated issues of resource depletion, war and climate change. Ignorance of environmental management and an apparent lack of planning for an egalitarian society with sustainable natural resources, means that another decline and fall of our collective civilization looks inevitable, unless we change our ways. Perhaps the adage about history repeating itself is the clue we need to up our game, and engage in meaningful, sustainable evolution that supports all human beings.

Each recurring ice-age produced huge drops in the level of the seas, as much water was deposited miles high, in the form of ice in the world's polar regions. This allowed people to wander ever further and to populate many places around the Mediterranean, even those which were later to become islands, with the return of higher sea levels during warmer times. No doubt as the climate warmed there were some spectacular extreme weather events that transformed the landscape, and some spectacular collapses in the rocky outcrops pushed up from Earth's crust by the shifting of the tectonic plates that brought the African land mass into slow collision with Europe. Precariously balanced mountains, obeying the laws of probability, sat there for millennia before the final straw that provided the momentum for gravity to transform relatively stable masses of rock, into moving cascades of rock and ice, creating events of mountain collapse on a grand scale both in space, and also spread out randomly over the time continuum. The tumultuous collapse was a series of cascading landslides on a grand scale, creating new rivers, valleys and the forms of the modern European landscape, from the Swiss Alps to the Mediterranean Sea.

MYTHIC REALITY

So whatever the original events that provided the mythic stories of Hercules, it is probable that they relate to journeys undertaken through regions that were changing dramatically in the wake of the end of the last ice age in Western Europe around 12,000 ago. What amazing people our ancestors must have been to have survived these cataclysmic events. They were human beings with the same brains and abilities as we have today. All of these events took place as our forebears were still in a grand phase of migration out of Africa, through a phase of development in the Middle East, and into post ice age Europe. Perhaps some of the stories that survived were of practical matters, such as changes in landscape and environment that affected traditional annual mass migrations of animals followed by our hunter gatherer ancestors. Seven thousand years after the end of the latest glaciation, the Bronze Age and the further development of domestic implements have left clues in the archaeological evidence. As well as pottery and clothing, the finds provide a view into the extent of travel of proto historic people worldwide, and an insight into the development of the pathways, transformed from seasonal wandering by clans in search of new territory and food sources, eventually becoming established as the earliest transport routes of what is now Europe. And the incredible truth is that people used legendary narrative, visual arts and rhythmic sounds to record their history, their world view, and even to describe their itineraries.

From around 3,000 BC there is archaeological evidence that a culture known as The Corded Ware, named for the characteristics of its pottery, spread across the Germanic territories, to northern Europe from the late Neolithic (Stone Age), throughout the Copper Age and culminating in the early Bronze Age. In parallel, the Bell Beaker culture was a similar phenomenon of people becoming increasingly mobile through the search for metals, and the use of horses for transport. The beaker pottery finds indicate this culture spread throughout Western Europe and across to Britain and Ireland. It was not just the particular pottery type, but it seems there was an evolutionary current for the search for minerals, and the practice of metallurgy that was part of a wave of migration, more a gallop than the previous slow moving population drifts. People had developed their equestrian skills, allowing rapid transit from one region to the next. They collaborated to develop metal working, and shared ideology and objectives characterised the rapid spread of this culture throughout Western Europe.

Similarly, the Golasecca and the Hallstatt cultures of the late Bronze Age indicate a level of mobility from our earlier ancestors on a scale that was missing throughout the modern era. The pressures of conquest, resource constraints in the dark ages, feudal, mediaeval and post mediaeval times up to the twentieth century, restricted and constricted migration. With the centralization of power by military conquest and the rise of princely states, people became increasingly tied to a single location, either by enslavement, or because of feudal overlords regulating land usage in the wake of militarism. The ability of hunter gatherer and early agrarian societies to adapt to changing environments by changing locations was over by the end of the Iron Age. Not till the nineteenth century would people again migrate en masse.

After the last glaciation, knowledge of navigation at sea and on land by our ancestors was developed, and reached great heights of sophistication, given that people were still at the point of developing technologies from available natural materials of stone and wood. The technologies and knowledge increased with the discovery of bronze metal working, and subsequently the smelting of iron, with its superior tensile properties. Heights that would not be scaled again until our present era.

Cultures such as the Lapita culture, a proto-historic Pacific Ocean people and society, dating from about 1600 BCE to 500 BCE have been skillful navigators for millennia. These people were antecedents of both the Polynesian and Melanesian nations, exercising wonderful navigation skills to find their way century by century, ever further across the Pacific, eventually reaching Hawaii. They originally set out from the Philippines in the centuries before our current era (BCE). Even today, the people of Torres Strait and other Pacific islands use a system of navigation to make voyages across long stretches of open ocean. Local knowledge of weather patterns, astronomy and seasons are still passed along by oral tradition, supported by music and dance patterns, from master navigators to the next generation of students.

Linguistic and DNA evidence indicate that the Lapita people who eventually made it to Hawaii, colonised New Zealand relatively recently around 1200 AD, during a favourable climatic period for prevailing winds. Navigation skills based on the motion of specific stars, where they would rise and set on the horizon of the ocean, the ability to read the waves, detect long range weather conditions, and understand the seasons of travel are still taught today. Knowledge of dugongs, turtles and whales and their habits are studied and understood. The direction, size, and speed of ocean waves, the exact colours of the sea and sky, and the positioning of clusters of clouds all provide location information for islands, as well as the correct angle of approach for safe landing on beaches and in harbours.

Our sea-going European ancestors would have had a much easier time navigating around the Mediterranean, it is true, able to sight landmarks as well as use astronomical positioning. It is only reasonable to suppose that people who travelled on land were using their intellects to develop similar skills for navigating through terrain that comprised featureless forests as well as confusing hilly and alpine regions. There are plenty of reasons to suppose that Celts developed sophisticated route maps based on the sun and the fixed stars, as well as solar and lunar timing for community events.

Is it really so surprising that Celtic peoples had sophisticated methods of navigation for land trading routes, only a few thousand years ago, clearly established over millennia? What is really interesting, is that according to recent research, Celts relied on local people to maintain the routes. As well as keeping the knowledge of the standard navigation measurements and locations, they noted the landmarks that provided directions, all based on seasonal variations of the rising of the sun, and their knowledge of the north-south and east-west axes. The co-operation between groups of people enabled them to provide input to travellers taking the long distance navigation trajectories, in a cooperative system of direction finding. There were probably established way stations for lodging en route. In the Celtic world, this collaboration was so successful because it was based on shared values and a common understanding of the technologies used for navigation. As people were initiated into the mysteries of the summer and winter solstice angles, measured during the week when the rising sun is at a standstill in summer, it must have been a wonderful festive atmosphere, providing a focal point for communication through stories and technologies, while providing a standard of navigation for travel and trade.

It would make sense if the Heraklean Way was originally part of the ancient route established after the end of the last ice age. The retreating ice allowed hunter gatherer groups to wander on the same paths taken by migratory animals, following relatively easy undulating routes across Europe, avoiding extreme and rocky terrain. By the time 10,000 years had elapsed from the end of the last glaciation, at the end of the Iron Age the Celts were able to use the solstice angles to codify the navigation of these routes, and communicate them from community to community, from one bounded tribal territory to the next. It seems reasonable to suppose that they developed these methods of navigation in a purposeful manner, in order to repeat what were originally meandering hunting tracks, to establish predictable and safe trade routes across proto Western Europe.

While originally these routes were along ways that provided ready sources of food and camping grounds, since the great melting of the ice after the last glaciation, they would have gradually made permanent settlements, inhabited during the summer months when work could have been done to establish shelters and dwellings with local materials. With the development of trade and agriculture in parallel with domestication of animals, came improvements in food production and social organisation of habitations. Technological progress was made for enabling navigation of land journeys covering long distances over vastly differing terrain, using the solstice angles, and right angled triangulation, sites for surveying, and landmarks for reference. It seems that our ancestors established the first system of long distance highways, as well as developing seaways to sail the Mediterranean, voyaging west to Hibernia and the British Isles. Meanwhile the seafarers of northern Europe were unfurling their sails in the higher latitudes.

The late Bronze Age collapse in the Eastern Mediterranean region, Anatolia and Egypt was probably the first of the large-scale declines and falls that have characterised civilization since this time. There were a number of factors that triggered the events leading to this wholesale breakdown in cultural, social and technology systems. The pattern seems to have been the combination of change from the impact of permanent agriculture on local ecosystems, and depletion of natural environmental resources, at the same time as increases in territorial incursions and warfare from the Eurasian Steppes. There was also a definite climate fluctuation that had a lasting effect apparently caused by seismic events, and exacerbated by volcanic activity in Iceland leading to a significant cooling of Northern Europe and decreased rainfall around the Mediterranean. These events would have caused population movements and migrations, including the phenomenon of 'sea people' moving into the Aegean and Egypt making use of sailing boats. The Bronze Age 'palace' cultures were replaced largely by village cultures. These movements would have affected both the trade routes and the population drift over the Alpine passes to Switzerland., During the centuries from 1150 - 950 BCE, when temperatures would have been much lower in winter, the population in the higher altitudes certainly would have declined because of the cold.

Two hundred million years ago, in the Triassic period, a giant supercontinent , Pangaea existed in the southern hemisphere. Asia separated and drifted north. Gondwana comprised all the southern landmasses, including Africa, Antarctica, Australia and South America. The continent of Australia-New Guinea began gradually to separate and move northwards 55 million years ago. Left over from the Gondwana continents are the rainforests. The fossil record indicates that when Gondwana existed it was covered by rainforests containing the same kinds of species which are living today in Australia. The evolution of the Earth's geology and biology is conserved in the remnant rainforests in Queensland, New South Wales and Tasmania.

In fact, a wide range of plant and animal species that are still living today have their lineages in ancient Gondwana, such as ferns, cycads and conifers. Rainforest once covered most of the great southern land. Few places on the planet contain so many plants and animals which remain relatively unchanged from their ancestors in the fossil record. There is a concentration of primitive plant families that are direct links with the evolution of flowering plants more than 100 million years ago. This preservation of evolutionary history is unique on the planet, and at the time of writing the Tarkine forests of Tasmania are still under threat from state sanctioned logging and mining.

Thermo-luminescence dating is the determination of elapsed time by emitted radiation once exposed to sunlight. People living in northern Australia have been dated to between 50,000 and 60,000 years ago, which accords approximately with the DNA evidence for migration out of Africa towards Indonesia, according to the National Geographic genographic project, a collaboration with IBM.

It has been suggested that the first arrivals in Australia were coastal people, basing their economies on the sea and river mouths, originally migrating along the coastline, subsequently heading up the rivers. If this is the case, since the mouths of the rivers were hundreds of kilometres seawards of the present coastline, the question then becomes how long did it take to reach the sites on present-day dry land? Presumably the first landing would have been on part of the continental shelf that is now submerged, an unknown number of years earlier than the known TL dated sites.

The first people arrived on the continent when sea levels were much lower than today, and were able to deal with the separation using boats, to island hop their way to Australia. Because of rising sea levels, there is no possibility of discovering exactly what kind of craft was used. However, there must have been enough people arriving, either by chance or by design, to begin a successful settlement in the new land. This migration via

Indonesia was thought to be achieved by a slow process, perhaps taking hundreds of years, although further to the east New Guinea, like Tasmania to the south, was still connected to the main continent by a land bridge at this stage. It is likely that there were separate waves of arrivals, and the earliest people may have been displaced by, or moved further because of the ingress of newer arrivals. There is a range of interpretations, one being that the fossil record of early Australians may indicate colonisation by two separate genetic lineages of modern humans.

DNA analysis indicates that the first Australians started their journey more than 60,000 years ago, branching off from humans who left Africa. The ancestors of contemporary Europeans and most other Asians probably went their separate ways less than 40,000 years ago. There is new evidence that indicates the possibility that the first Australians may have a small percentage of additional DNA, not only of Neanderthal, but also another homo erectus from Asia in their lineage, which means that they may have more biodiversity in their genetic material. Human populations were so nearly wiped out so many times since the waves of glaciation in the Pleistocene epoch, lasting from 110,000 years ago and ending about 15,000 years ago, that today we have significantly less biodiversity in our genetic code than our closest cousins, the chimpanzees.

There is no doubt that the arrival of people in Australia meant dramatic change for the plants and animals living on the continent, with the recorded disappearance of the megafauna, the giant goannas, kangaroos and emus, around 45,000 years ago. With these species gone, the entire spectrum of plants that were part of their food chain also had to evolve or perish. Human beings and our propensity for hunting also managed to extinguish the European megafauna, the mammoth, albeit somewhat later, around 5,000 years ago. The moral, perhaps, is that as a species, when we develop a capability, we do not always know when to stop. The successful strategies of trapping megafauna by using geographic features such as cliffs and ravines, valleys and impasses from which they would be unable to escape did provide an abundance of fresh meat in the short term, however a round of extinctions of the major source of protein must have meant a significant disruption of lifestyle for our hunter gatherer ancestors, and the need for rapid acquisition of new skills for hunting different species.

Fast forward to the Bronze Age in Europe and Asia, around 3,000 BCE, and it seems that human activities were changing the environment of the temperate grasslands that were once the northern Eurasian steppes, extending over a vast region from the Danube almost to the Pacific Ocean. Responding to changes in the ecology, people developed a more mobile society with corresponding economic activities, moving with their herds of cattle in a form of nomadic pastoralism, in which the development of textiles for domestic use became increasingly important. Domestication of the horse, and the development of wheeled vehicles in the northern Caucasus, the Iranian plateau, and the ancient civilizations of Iraq, changed the course of European history.

The 'Fertile Crescent', in ancient times the region around the Nile, Tigris and Euphrates river deltas that allowed ancient Mesopotamia, Assyria, Phoenicia and Egypt to evolve, was home to early agriculture. Four of the five most important species of domesticated animals were cows, goats, sheep, and pigs, and a fifth species, the horse, lived nearby in the Caucasus. However, lack of knowledge of exploitation and conservation of topsoils, the importance of trees and wild plant species, and the maintenance of water sources meant that eventually, the local climate destabilised, river courses changed, and productivity could only be maintained with the extensive use of irrigation causing widespread salination.

Meanwhile, in Australia, cut off by geography after the closure of the land bridge to Asia, lifestyle continued in the traditional lineage of hunting, gathering, and long distance communications for trade and ceremony. Most importantly, the first Australians continued to develop the large scale land management practices, as a response to changing environment, harsh climatic conditions, fragile topsoils and low rainfalls. It is only in the past 200 years that Australia's land management practices have become seriously imbalanced. Sadly, in Australia in the 19th and 20th centuries, European farming methods were inappropriately applied to areas with inherently fragile top soils, and a delicate wide area ecological balance, completely unknown in the parts of Western Europe from which Australia's new arrivals came. The result is that large areas of land are now seriously degraded, overrun with introduced species of plants, insects and animals that have decimated native populations. The rate of environmental degradation of local terrain and waterways is universal, ubiquitous, and the percentage of unsustainable agriculture, silviculture and aquaculture increased dramatically during the past thirty years. It is only in the past ten years that ecologically sound methods of agribusiness have started to make their way into the mainstream farming practices.

Despite the development of wide scale custodianship of their ancestral 'country', and a peaceful existence for the most part between tribes in the past millennia, it appears Australian indigenous culture also suffered at times from tribal hostility and territory conflicts. When the British arrived in 1788 it was a time of trade and allegiance of tribes through marriage. There were strict codes for marriage 'skin' groups, resource management, environmental conservation, and punishments for settling disputes between people, tribes, and banishment for marrying 'wrong skin' (violating laws developed to inhibit overpopulation). As far as can be determined, no one tribe conquered vast swathes of territory, nor monopolised the wealth, nor enslaved other people to any degree.

In Australia there is a rich history of art depicting people who were present over 50,000 years ago living amongst the megafauna and exotic plant species, hunting and travelling seasonally for food sources. The Yalata people from the Nullarbor Plain in South Australia still relate the oral history of events caused by the end of the last ice age when from 18,000 to 5,000 years ago, melting polar ice sheets caused sea levels to rise, inundating the shoreline which moved several hundred kilometres inland, roughly where it is today. Like the first Australians have done for thousands of years, people still gather to watch the migration of southern-right whales which arrive from Antarctica between June and October.

Meanwhile, In Europe, 17,500 years ago, bands of hunter gatherers were also wandering in search of game and plants for food and medicine, drawing equally stunning depictions in ochre of the flora and fauna in caves such as at Lascaux, as did the indigenous Australians, their counterparts on the continent of Gondwana, the great south land.

The Middle East was one of the key locations in the development of farming and agriculture. Hunter-gatherer lifestyles of modern humans were replaced in Europe with permanent settlements with a reliable food supply. Accompanying this fundamental shift in human evolution, the global population began the explosion that means the planet which sustained around five million people 10,000 years ago, now supports more than seven billion. Analysis of finds suggest that the first domestication of wheat commenced in a small region of south-eastern Turkey, from whence it spread to the people who lived around the eastern Mediterranean region, including Egypt, Iran and Jordan.

This discovery of wheat as a crop was caused by the availability of suitable grasses. Wheat's ability to self-pollinate greatly facilitated natural selection and domestication as a crop. This discovery was probably assisted and supported by a period of climate change, the Atlantic, or Holocene Climate Optimum from about 8,000 to 5,000 BCE, when temperatures in the Northern Hemisphere were warmer than average during the summers, and the tropics and areas of the Southern Hemisphere were colder than average, with a sea level around three metres higher than today in northern Europe. The combination of the use of stone axes, fire and grazing animals to clear land that was used for cultivation of crops, began to change the ecosystems forever.

Indo Europeans made their way into the Alps after the last glaciation retreat around 12,500 years ago. Perhaps some earlier traces still existed, of past people who had lived in the Alps prior to the Rhone glaciation 30,000 years ago, although the flora and fauna had been engulfed by the rivers of ice. Perhaps as they roamed and hunted they found fragments of evidence of past inhabitants, flints in the tumble down scree of the terminal moraine of the glacier, remnants of antlers and bones, and other echoes of a far earlier people.

Since the retreat of the ice in Europe, around the time of the Great Flood in Australia, gradually people would have formed tribal loyalties and allegiances, and slowly established travel paths, routes and places for ceremony, culture, trade, marriage, language and exchange of information. This was the pattern of these people, who had the same physical capabilities, constraints and characteristics as ourselves, with no evolutionary biological difference from ourselves. In fact, the survival skills of every individual, required prior to the development of towns and cities, meant that only the ablest, best adapted or luckiest individuals did survive in those times. The DNA of those human beings is no different from our own, and a baby of today would equally fit into stone age existence as a stone age baby could adapt to a modern lifestyle.

The European grasses were amenable and tractable to agriculture, so farming settlements became feasible in the past 12,000 years, as seasonal changes after the last ice age favoured wild cereals. The more fragile ecosystems in Australia, after the Great Drought, 30,000 to 15,000 years ago, with vast tracts of desertification, meant that firestick farming prevailed. Until two hundred years ago, indigenous Australians practised large scale land management over the taming of the seasonal output of the land over vast tracts of country, using the burning off and regrowth cycles, ensuring that people were better off maintaining a nomadic lifestyle.

Similarly, in Europe, the small groups of hunter gatherers gradually increased in population and became localised tribal groups. As agriculture enabled semi-permanent settlements, location became an increasingly important asset. In heavily forested regions, seasonal migration was still a more important factor than social organisation around a township. Maintenance of temporary and permanent shelters became a focus, and this entailed local co-operation. Eventually, by more or less mutual agreement, tribes acquired fixed location boundaries, and promoted a certain style of co-operation for mutual benefit. Respect for traditional land and migration paths enabled groups of Gallic peoples, in particular, to form a culture, which, like the practices of their indigenous Australian counterparts, garnered and propagated local lore, knowledge, law and ritual, sharing their culture in a manner that owed more to marriage, trade and the ability to learn languages, than to warfare.

Development of statehood based on mutual advantage is a strong feature of European proto-cultures. In parallel, in those societies where people gathered closely together in towns, with specialist roles, losing their essential day-to-day connection to the land, life was subject to dramatic swings, as civilizations rose and fell. The decline and fall has been attributed to some common factors, mostly overuse of ecosystems and environmental resources leading to a collapse in the essentials for survival, over-ambitious expansion of territory and power, as well as territorial incursions by people in search of new resources.

Reading from various sources the myths and legends around the Greek hero Hercules, it is only one ancient Greek writer, who claimed to be Syrian who equated the Celtic god Ogmios with Hercules. In fact, the interpretations of the labours of Hercules are all by late writers from Greek and Roman times. The original story appears to have become mythologized and made heroic in the telling over thousands of years. For anyone familiar with indigenous Australian dreamtime stories, there are resonances with the tales of Hercules. Australian dreamtime stories generally provided explanations of geological and geographic features and history, told through the interaction of mythical creatures and the people. Ancient Australian creation stories provide rich meaning of historical events, and 'dreaming' often refers to the energies associated with animals that inhabit the local terrain. There are no stories with a warrior hero and a beautiful maiden as central characters, something that appears frequently in European and Middle Eastern mythic tales.

The proto-historic Australian narrative encompasses a range of characters, men, women and animals, who all play a role in the creation and maintenance of terrain, food and water sources, and cultural practices and law. Some of the deepest stories are related to the forming of geographical features and locations, clearly aimed at retaining a history of the changing landscapes, climate and major geophysical events over the 60,000 or so years since the first Australians arrived by boat to the great south land.

Indigenous Australian creation stories can provide a compelling comparative parallel to Aegean civilization myths and legends. It is beginning to be understood that allegory and metaphor are a standard way for humans to perpetuate important historical events, which invoke deep psychic responses in the human condition. The stories may gain in the telling, and change shape during the oral narration over centuries and generations, however there always seems to be a coherent thread at the heart of the story that crystallises over time. First Australians are, today, still telling the narrative of when the coastline came inland 200 kilometres at the end of the last ice age. They have been telling variations of this story for the past eighteen thousand years, since melting polar ice began drowning a quarter of Greater Australia.

The Celts are not the only people who appear to have been linked ethno-linguistically as a group of tribal societies. It is clear that the first Australians, with different language groups, and varied cultural practices, also had trade routes and cultural ceremonies that linked them together. Some of those people travelled over a thousand kilometres on a seasonal and solar annual basis. From as early as thirty thousand years ago evidence survives today on prominent landmarks, recording the location of water sources, plants and animals, geographic features such as permanent springs, and valleys joining them, along the seasonal roaming and trading routes of the

First Australians.

Australian aboriginal peoples have been undertaking nomadic journeys over many hundreds of kilometres over at least the past 50,000 years according to archaeological evidence from cave paintings, when Australia was still connected by a land bridge to New Guinea and parts of south eastern Asia. The European myths may well, on deeper examination, provide a record of ancient narratives told around the fires of people whose primary concern would have been survival, food, shelter, and economic opportunity to be gained from occupation of new territory. Knowledge of local geology and geography, food sources and water were everyone's business before the rise of the 'palace' states. The people of proto history must have been more concerned with gathering together for mutual support and survival, before the population pressures and climate change in north Africa meant that competition for scarce resources changed the modus operandi of our ancestors.

Prior to these times, people must have been engaged in a compelling and all absorbing lifestyle of finding food and shelter. Of course they must have had aspirations for quality of existence. They lived, through necessity, more or less in harmony and balance with their environmental conditions. The vicissitudes of nature may have been their primary concern, but that did not stop them striving for understanding of the universe around them, to better their living conditions, when it was provident and prudent to cooperate for survival. The ritual of the campfire and the need for hunter gatherers to provide mutual protection would have been sufficient incentive to swap stories, develop allegory, metaphor and myth, just like their counterparts in Australia. The development of agriculture and settlement, and the increasing pressure of population on arable land would start an entirely new era of human development in Europe and Asia.

Narratives of Hercules provide an insight into the concerns of our ancestors in the millennium before our current era, when conquest and territorial wars were on the increase. There appears to be a parallel rise in the cultural status quo of the warrior, at least in some parts of the world. This includes Celtic society, where women were active participants in all spheres of social life, including the defence of home and territory, and the inheritance of property. This change in our collective psyche provides another indication that heroic myths reflect the cultural and socio-economic concerns of the times. The stories of our ancestors echo their hopes and fears, generated by the growing challenges of population expansion, the finite availability of natural resources, and the increase in the level of territorial incursions into the agrarian settlements.

GREAT SOUTHERN LAND

The events of European prehistory cannot in all conscience be classified into a neat timeline with clear delineation across geo-political boundaries. One thing we can be sure of is that there was just as much wrangling and negotiation to determine actions by a clan group as there is today. Societies did not just stop being hunter gatherers when they learned to cultivate wheat, nor did they become exclusively wheat or cattle farmers. The semi-nomadic lifestyle of our ancestors would have gradually evolved over thousands of years, until climate and terrain determined a practical mix of food harvesting activities. There are distinct regional histories. For example, people in the pre-alpine regions throughout Europe continued to graze their cattle in summer pastures, returning to warmer places at a lower altitude, establishing temporary camps, a lifestyle that continued until the 20th century.

Radiocarbon dating has accurately established that there was a distinct gradual communication of agriculture practices out from the region of origin in the Middle East.

The waves of advance of people westward is indicated by a slow propagation of agricultural methods at a pace of around five hundred kilometres every five hundred years, which, at one kilometre per year, is a rate of change commensurate with population movement between 8,000 and 5,000 years ago. This slow pace of propagation belies the evidence of journeys between localities, as evidenced by archaeological finds from one region found in terrain hundreds of kilometres distant. People were travelling in search of opportunity, but it cannot have been easy.

During this same period in Australia, a very fragile climate meant that much larger areas of land were required to support a lower population density than in Europe, with its forests and grassland. Correspondingly the first Australians continued to develop lifestyle improvements based around seasonal migration, including large scale firestick farming to maximise the food production potential of their lands. From around 7,000 BCE people in Australia learnt to manipulate available plants and animals to increase food resources on a grand scale. They transformed an entire continent into a fully sustainable estate, until the arrival of British and eventually other European settlers, destabilised the environmental balance achieved over 50,000 years. This was because of the application of farming methods from entirely different climatic conditions.

Recognition of the requirement to adapt to different ecosystems came slowly in the great southern continent, after sequential environment and economic failures in the 19th and 20th centuries, caused by disturbing the interaction between the land, the rains, the river systems and increasing soil depletion and salinity. These problems have only started to be seriously addressed in the 21st century. It is safe to say that our knowledge of the balance between nature and human society is still insufficient. One interesting hypothesis is that greenhouse-gas emissions produced by early agricultural activities, and the burning and clearing of forests in the past couple of thousand years has kept the climate warmer than its natural level, and possibly offset an incipient glaciation.

The driving force for all humans has always been the betterment of living conditions through innovation using local materials, food produce, and management of flora, fauna and environment. Trade of goods and materials has always been a method for optimising the potential for economic improvements, and has always worked best in times of peace and sustainability of resources. The first Australians had laws and methods of governance to ensure that people complied with the law, to maintain local ecosystems, which included a limit on population achieved by a strict marriage code. The Central Land Council of Australia records: "The kinship system is a feature of Aboriginal social organisation and family relationships across Central Australia. It is a complex system that determines how people relate to each other and their roles, responsibilities and obligations in relation to one another, ceremonial business and land. The kinship system determines who marries who, ceremonial relationships, funeral roles and behaviour patterns with other kin."

It seems that a more fragile set of ecosystems in Australia and continuation of hunter gatherer nomadic lifestyles culturally encouraged behaviours that enshrined above all, respect and maintenance of the sustainability of the natural environment. This respect is mirrored in more rural, remote mountainous and desert regions in Europe and Asia, but was nearly lost in towns and villages in the last two millennia. The most likely causes were the loss of the right to roam after the Roman era, the growth in population that accompanied settled agriculture, and the developments of urban settlements (decreasing environmental sensitivity).

For the most part, Australian lands were never suitable for intensive cultivation, and large tracts of land, particularly around the Murray River, have been turned into saltpans by the European farming practices of the past 200 years. Since the British declaration of 'terra nullius' and the claim of the continent for the British crown and establishment of a harsh penal colony at the end of the 18th century, species invasion has been the name of the game, not only in Australia, but echoing the events of European colonisation in Africa, Asia and the Americas. This historical period was part of several centuries of territorial expansion with an arrogant mindset, which has resulted in the widespread displacement and dispossession of traditional people the world over, and the associated downgrading of the planet's ecology in the name of short-term profits.

Abundance of money resulted for comparatively few families who for the most part never set foot on, and certainly did not provide the labour for the lands that provided them with riches. Mostly, the hard work in the fields, the mines, and the transport of goods was either provided by slavery or at a level of remuneration that was extremely low in comparison to the profits. Labour conditions were appalling in Europe as well as the European colonies. It took the next two centuries until the end of the twentieth century for even Western European peoples to gain a living wage, and enjoy a good standard of living from profits generated from colonial expansionism.

Probably for this reason there was no widespread dark age in Australia such as the one that succeeded the fall of the Roman Empire in Europe. Despite over two hundred years of European conquest of Australia, bringing genocide and deprivation to the indigenous people, there is an unbroken continuity of a culture based on traditional custodianship of land and resources essential to life, particularly water, and habitat for all wildlife. Respect for the animals, particularly those that were hunted for food, was a natural state of affairs that occurred in all civilizations keeping close ties to the land, and the environment that sustained the people over the centuries.

The first Australians are still in touch with the traditional knowledge of their ancestors, in an unbroken line over the past 50,000 years, and Australian aboriginal society is enthusiastically embracing the age of information and communications technology, having largely skipped the industrial and post-industrial eras. Their knowledge, experience and understanding of living in harmony with the seasons, and grand scale land management, as custodians of precious natural resources, is today providing valuable lessons and education for the sustainable development of Australian agribusiness.

While Asian nations struggle to catch up with western wages and conditions, multinational corporations' super profits soar into the stratosphere. They are, however, just as at the height of European colonialism, distributed only to a very few. It does not take a genius to figure out that this situation must inevitably lead to widespread political and social unrest unless addressed urgently. In the first part of the 21st century, many governments in all parts of the world lurched to the right, encouraging conformity, decreasing spending on the social conditions (so hard fought for in the twentieth century), while providing no discernable leadership in regulating the response to sustainable innovation with technology.

It is not the wealthy corporations addressing climate change, even though most of the greenhouse gas emissions have been caused by under 100 companies.

Many are currently conducting ad campaigns and PR exercises on a grand scale to boast their very minimal green credentials. The United Nations, despite its best efforts to promote social justice, cannot force corporations to conform to decent and responsible behaviours, and it remains to be seen whether a grassroots movement can provide the social influence for concerted action to reduce global warming, in the face of national and corporate apathy.

The good news is that globally and collectively, we have all the technologies, methods, knowledge and best practices at hand that can put our unsustainable ecosystems to rights with a globally collaborative effort from people of every nation, addressing problems on every continent. Hopefully the Arctic and Antarctica can still be preserved from wholesale economic exploitation. Land can be self-sustaining with strategic design and a minimum of intervention by using local knowledge of the natural terrain, and whole system environmental management. This includes forsaking inappropriate farming methods. Today, application of appropriate knowledge and technology means that the use of toxic pesticides is no longer necessary, something which is being resisted by many governments and agribusinesses. They have a lot to learn from indigenous Australian practice by a people who for the past 50,000 years at least, developed ingenious systems of sustainable land, river and ocean management, to foster natural resources for the good of all for the long-term.

THE UPPER RHONE VALLEY

During the last ice age, a gigantic glacier cleaved the alpine rock aside with the ineffable power that water has on every other element. Once the temperatures allowed the alpine glaciers to melt, the process had an incredible effect on the terrain in Switzerland, and also the surrounding landscapes of modern Italy and France, as the enormous pressure of the ice, sustaining the rocky outcrops of the mountain massifs, was gradually released. Gravity would have played a part as rocks, propped in place by gigantic blocks of ice, rolled from points of higher elevation, gathering momentum and debris as the mountains were released from the restraints of the ice. The last great glacier retreats in the Alps started around 10,000 years ago, and the melting of the glacier nearly half a kilometre high in places, left the lakes Neuchatel and Leman (the lake of Geneva) in their wake. Large granite boulders can be seen scattered in the forests of the region.

The retreat was so rapid that within two or three centuries, the glaciers became miniatures of their former selves, locked into the highest Alps. During the melt period, a huge lake flooded the region to a depth of several hundred metres before levels fell, leaving traces in the hills at the foot of the mountains, and forming the gentler slopes and small foothill plateaux over glacial moraines. Coupled with the continuing movement of the tectonic plates from the collision of the European and African continents, a combination of seismic instability, gravity, and ongoing changes in the Earth's mantle, the Alps continue to morph and reshape with rockfalls up to the present day. With the effects of global warming accelerating the melt rate of the remaining glaciers, further movements can be expected.

History and archaeological records show that since the Bronze Age, the Chablais Alps, particularly Les Dents Du Midi have been particularly unstable, with massive collapses occurring several times, accurately dated through both remnant vegetation on the bed of the lake, and through historical accounts. The last rockfall, though minor in comparison to the large seismic events, was significant. A mass of around one million cubic metres of rock detached from the side of the Haute Cime and fell to an altitude of about 3,000 metres, fortunately at a higher altitude than any human habitation. So what is the pattern of migration since trees started to grow again in Switzerland, and what part did climate play in settling the Upper Rhone Plain since the last glaciation?

During the development of agricultural practices, once

the turbulent cataclysms of melting ice, rockfalls and mountain collapses, new rivers from glacier meltwater flooding valleys, and seismic tremors ceased, our ancestors began to repopulate the Swiss plateau from Gallic, Ligurian and Germanic tribes. It is likely that these people were originally alpine cattle herders grazing their animals, in the mountain regions surrounding the Alps. They would have lived in small defensible settlements growing crops, and grazing their cattle in mountainous summer pastures. They would have migrated as clans, founding wayside campsites and shelters as they ranged further in pursuit of more land for grazing and seasonal farming. As the northern hemisphere warmed, people would have found their way into the Upper Rhone Valley, and as in other regions of Europe, our ancestors gradually self-organised into villages and townships, with shared labour, skills and knowledge throughout the Neolithic, Bronze and Iron Ages.

According to Swiss archaeologist Carine Wagner, evidence indicates that farming and food production have been practised since Neolithic times in the Chablais region at the head of Lake Leman, the lake of Geneva. The name comes from the Celtic Pennoslacos, (later latinized to caput laci, head of the lake), and it refers to the fertile alluvial terrain between the head of the lake of Geneva extending to the narrow passage where the Rhone passes between mountains at Saint Maurice, which in Celtic times was guarded by the Nantuaten oppidum, Tarnaias.

This implies that once the glaciers melted, people were

quick to seek out new territory on the Swiss plateau, using the available migration routes east from modern day France, north-west from Italy over alpine passes, and southward from Germany. The abundance of Neolithic finds of pottery, implements and natural materials, from hundreds of kilometres distant, indicates that trade and culture were indeed being propagated widely by travellers at an early stage of history. There were clearly established paths over the Alps, and it seems that people were just as interested, as we are today, in learning new skills, and acquiring new technologies for use of raw materials, over increasingly long distance trade routes.

So what was happening to the ancestors in this flood plain where the Rhone meets the head of Lake Leman, at the opposite end from Geneva, during pre-history? In brief, because of the glaciation, there is little earlier evidence of humans in the Alps. However stone and flint tools dating back to 35,000 BC were found in a cave above the French side of the Rhone plain. Bone relics of numerous animals from around 12,000 to 13,000 years ago, after the retreat of the glacier, date the existence of a band of hunters who established themselves around the head of the lake. They were replaced by Magdalenian hunter-gatherers of Western Europe after red deer, horse and other large mammals. They roamed seasonally, perhaps had proto trade routes, and were thought to be related to the people of south western France.

After the end of the last Ice Age, from 9,000 BCE, Mesolithic hunters colonised the valley in two ways, from the high altitude passes from Italy, and along the lower Rhone valley from France. The Neolithic agro-pastoral economy was connected via the passes from the Alps to the Po plain, possibly related to the practice of summer pastures for small livestock at higher altitude. Between 9,000 and 6,000 BCE people lived in stone shelters, and lived off the plain and the wooded slopes above it, hunting cerf, bears, wild cats, tortoises, birds, gathering bird's eggs and nuts, clothed in animal furs. Their tools were mainly finely worked flint gathered from the mountain massifs, such as Les Dents du Midi. Perforated shells from the Mediterranean have been found in the region, evidence that people were roaming far and wide.

The end of the ice age meant that trees once again flourished in Switzerland. The upper reaches of the Rhone flow through what is today a narrow plain bounded by rising mountains, where once there was a vast river of ice. Cultivation techniques, the spread of the invention of the wheel, and the establishment of settlements on this fertile plain commenced in earnest around 5,000 BCE, as people followed migration paths from southern France along the Rhone river, and northern Italy over the alpine passes to reach the lands around the shores of Lake Leman, and the plain of the vast valley carved out by the glacier.

The early Neolithic period from around 5,500 BC was characterised by a superior use of tools, enabling a more settled existence than that of the nomadic hunter gatherers. Comparatively sophisticated stone axes, allowed timber to be felled for houses. Clay started to be fired and cooking pots promoted a more varied cuisine. Evidence from around 5,000 BCE indicates that agriculture commenced around this time, at least at Sion, and that groups of people wandered north from Italy through the Alpine passes, suggested by the similarity of cooking utensils to those found on the Po river plain, south of Venice, attesting to the initiatives of people to move in search of new opportunity.

During the late Neolithic period, (around 4,500 to 3,500 BCE) there is evidence that the Chablais was inhabited, particularly at sites at St Triphon, a few kilometres from Aigle. At these locations, there are low foothills under the towering mountains, which would have been attractive to settlers, as the Chablais is basically a flood plain. Until the taming of the Rhone by building of channels and dikes in modern times, the land around the river would have been marshy, unstable, and prone to being inundated. Finds from this period show domestication of goats and sheep, and artefacts in glass and polished stone, though the hunt was still very important. The dead were buried in stone edged graves, and the lower hilltops were places of significance to the locals, perhaps as gathering places, perhaps for ceremony and communication. The Rhone plain would have been a relatively predictable and safe environment, particularly in comparison with the surrounding wild wooded valleys that led to the towering Alps.

Downstream at the lake delta, before banks were

constructed, the Rhone would have meandered through the land at the head of the lake, sometimes flooding the alluvial sandbanks. Along the shores of Lake Leman (Lake of Geneva), lake dwellers had established habitations around 6,500 years ago in the Neolithic and Bronze ages. Settlements of raised stilt houses became common around Alpine lakes. Discoveries of a number of these proto villages were made in the nineteenth century on the lake shore at Geneva and at Morges near Lausanne. The preservation of organic material, the remains of the wooden posts which supported the houses, deposited in waterlogged ground, provide unique sources of information on the history of these early agrarian societies. Research has determined that there were successive waves of 'pile' house buildings that stood in sandy soils at the lake's edge. Evidence points to five separate building periods, dating from around 4,500 to 900 BCE. People in these settlements engaged in basic forms of agriculture, growing food in the available land between lake and forest, and keeping livestock.

Woodworking tools of stone, antlers and bone were used extensively. From around 3,000 BCE, plough shares for cultivation of a range of crops such as wheat, barley and millet were developed from the Neolithic to the Iron Age around the lake. As well as bones of domesticated cattle, pigs, goats and sheep, game such as deer and wild boar show that people continued to hunt, either to supplement food or as a staple, during this period. A striking feature was the level of woodworking skill used for houses, pathways and palisades of the lake dwellers. Clearly the use of wood was an adaptive development suited to the local environment, and the 'pile dwellings' worked well on the lake's edge, protecting people from seasonal inundation. During the time up to the late Bronze Age, the surrounding mountains were densely wooded, and populated with bears, wolves and lynxes. The lacustrine people used the narrow strips of land near the lake for cultivation.

Still water in mountain lakes permits very fine particles of sand, silt, and clay to settle and form lacustrine deposits, exposed by changes in water levels. Despite the presence of alluvial silt, there is no evidence of any lake settlement at the head of the lake in the Chablais, although Lake Leman is a large freshwater basin formed by glacier meltwater. Perhaps the layers of soil in the delta meant that the land was too unstable. Perhaps the timber for the pile dwellings would have to be felled and transported longer distances through marshy ground, from the edge of the forested slopes around the valley, and this may have been a disincentive. It may have been that the shifting channels and sands of the Rhone, a fast-flowing torrent at the delta where it flows into the lake, were just not suitable for construction of dwellings. At this point, it is a great mountain river having collected waters from the valleys of the Alpes Poeninae to the south, as well as the Druse River at Martigny.

Another factor may have been the separate events of large scale mountain collapses near the river delta at the head of the lake, due to seismic activity during the Bronze Age. Dendrochronology enables the accurate dating of these houses on stilts near Lausanne, and there is a clear human occupation gap that may have coincided with the mountain fall around Les Dents du Midi and Grammont, from seismic activity that triggered tsunamis along the shores of the lake in the Bronze Age. What is known, based on sediment and geophysical analysis, is that large amounts of sediment from lateral slopes were transported and deposited underwater to the deep basins of the lake at this time.

In the Bronze Age people began to move out and up from the Rhone Plain in the Chablais, cautiously edging into the surrounding forests that rose more and less steeply from the flood plain. There were three distinct regional groups, the earliest of which was established in western Valais at the end of the valley furthest from the lake. No longer was this corridor along the Rhone just a migration route over the Alps, but an area with its own character and its own culture and style of development. Early Bronze Age finds dating from around 2,000 BCE, include tombs containing various artefacts such as bracelets, spirals, pins and diadems, mostly around St Triphon. Termed the 'classic phase of the civilization of the Rhone', the objects, weapons and ornaments, are much more sophisticated than earlier artefacts. The archaeology indicates that trade with Italy and other parts of Europe was already well established, and the Upper Rhone Plain by this time was on a flourishing trade route from Italy over the Alps, along the route that had been in use for population migration since shortly after the end of the last glaciation. Artefacts from the Po Plain in northern Italy have been found at St Triphon, providing evidence of early transience over the Alps. Settlements began to be surrounded by defensible stone walls during the Bronze Age. The flat plain land between the mountains on either side of the river, was used for grazing the herds of adjoining settlements. This site would have been a strategic holding for the occupants throughout the Bronze and Iron Ages, located on the only comparatively wide valley through the surrounding Alps, just a few kilometres from the head of Lake Leman.

The rocky headlands of St Triphon jut out from the glaciated plain, isolated both in vegetation and location from the surrounding hills. St Triphon is unique on the plain as it is an overturned nappe, a large sheetlike body of rock above a thrust fault, pushed along a few kilometres from its original position by enormous tectonic forces. Nappes form where there is a continental collision zone, where tectonic plates move over each other. This is doubtless part of the instability that has caused large rock falls from nearby Dents du Midi. The term stems from the French word for tablecloth, for the rock movement resembling a crumpled cloth being pushed across a table. Over thousands of years, large scale rock falls from neighbouring mountains, such as the one that fell on Tauredunum in 563, have occurred. The St Triphon nappe is also responsible for the large deposits of salt at nearby Bex. Quarries in the cliffs of the hilly outcrops of St Triphon have been exploited since the Neolithic period for the extraction of limestone.

As the late Bronze Age gave way to the early Iron Age

from around 800 BCE there is little to show in the way of archaeological evidence in the Chablais. Perhaps the population became sparser, in accord with the late Bronze Age collapse of the palatial civilizations in the Aegean, Egyptian and Anatolian regions due to a dry spell. There was also a decline in temperature in Northern Europe at this time, and because of its altitude, the Chablais region probably became more sparsely populated, due to the onset of a colder climate. The next clear evidence appears around 600 - 500 BCE. Fortified establishments around administrative centres were surrounded by tombs rich with objects such as weapons, jewellery and vases, from a trading class who were transporting metals, wine and other goods. The alpine passes became viable commercial trade routes. This was a very busy time of exchange of goods including bronze and ceramic vessels, and amphorae of wine imported from Italy, Greece and Marseilles, part of the emergence of Celtic trade. Ornaments found locally are like those found in the Mediterranean, Eastern Europe, and Italy, indicating that once again, the alpine passes were part of a highway that stretched from the Aegean to Britain, overland through the Upper Rhone Valley, as well as the better known sea routes through the Mediterranean, the Atlantic Ocean and the North Sea.

A little further upstream, Martigny is located in the middle part of the valley originally carved out by the Rhone glacier. It must have been a great relief to travellers after an arduous journey over the steep and challenging Great Saint Bernard alpine pass to reach the open spaces and level ground of the alluvial plain at last. Martigny, known in Celtic times as 'Octodurum', was a staging post, one that could not be avoided by travellers coming over the Alps through either the Great Saint Bernard or the Simplon passes, en route to Gaul (in extent largely equivalent to modern France). Martigny, located where the Upper Rhone Valley turns through ninety degrees, has always been a major crossroads for travellers. As well as the alpine passes to and from Italy, there is a route through to the Chamonix Valley, the shortest direct route to Geneva. And for those headed for ancient Gaul, the way continues along the Rhone through the Chablais to the head of the lake. One track diverged via the south side of the lake, an alternative route to Geneva (via Evian). Another headed towards Vevey on the opposite shore of the lake, then northward to the Celtic treasure trove Mormont. From this point travellers either ventured towards Dijon, and onward to the French coast en route for Britain, or branched along the Swiss Plateau to Neuchatel, and on to Germany. Little wonder that this region became strategically important, not only to the Celtic people, but as trade developed, to the Romans who were enviously observing the Celtic traders' commercial successes.

People had to have been innovative and inventive to carve out a niche and a lifestyle in this region, surrounded by the majestic mountains. As early as the bronze age, land would have been at a premium, as the Chablais plain around the Rhone where it approaches the lake only comprises around 20 square kilometres, much of it marshy flood plain, restricting its use for cultivation. Clearly reliance on crop cultivation would have been a risky business. Animal herding and hunting on the slopes above the plain had to be equally important for food supply. Thus, the economy must have been a mixture of influences. Evolution of the earlier lakeside village settlements would have been made possible by the influences crossing the alpine passes, bringing new forms of pottery, metalworking technology, grains and grapes, oil and fine art ornaments.

As with everywhere else in Europe, at first tools of copper, then of bronze (an alloy of copper and tin) were increasingly used as replacements for stone tools, providing improvements for plant cultivation with new agricultural implements such as scythes and sickles. Metal fishing hooks also appeared on the scene. The Bronze Age was characterised by a high population growth because of better food production enabled by metal tools. There is not a large amount of information connecting us with our ancestors during this period of history on the Upper Rhone plain. Certainly the standard of living would have been raised by superior technologies and tools such as metal axes. Improved methods of food production and livestock management would have enabled people to make better use of their time. The all-important survival necessities, food and water, were in abundance. Shelter construction from advances in woodworking was making life more comfortable. Knowledge of furnace temperatures and conditions for smelting would have been acquired, and become the subject of experimentation for successful metal working.

Iron production is cited as beginning in Anatolia around 1200 BC, though some archaeological evidence points to earlier use of iron smelting. Recent discoveries point to the development of iron working in several locations, including India and Africa, rather than dissemination solely from the Iranian plateau. During proto history, the time scales for the spread of new technologies was very slow, accompanying the advance in population migration and the development of trade routes. The spread of iron working west towards the Swiss plateau is regarded as taking place somewhere around 800 BCE. While trade routes developed and flourished, people continued to engage in the traditional semi-nomadic existence of early agrarian times, migrating over all the Alpine passes from Italy and France.

Iron must have been a great blessing to our ancestors. The ability to fell trees, and dress timber would have made the houses more comfortable and weather proof. The axe would have gradually changed the lower slopes adjoining the Rhone plain. People began to clear land for different farming practices and forestry management. Perhaps it was at this period that grape vines began to be grown in neat rows on the hillsides adjacent to the plain. These developments were in accord with European Celtic culture, broadly known as La Tene, because of the discovery in 1857 of rich archaeological finds at La Tene on Lake Neuchatel, lasting from around 450 BCE to the Roman occupation. The artefacts indicate a lifestyle rich in metalworking, geometric symbolism and artistry, building and navigational skills and advanced social structures, typical of the European Celts. Lake Neuchatel, like the neighbouring Lake Leman, was formed after the last glaciation, and the Swiss plateau was an important trade route of the Iron Age Gallic and Helvetic Celts.

By the time of the late Iron Age around 450 BCE in the Upper Rhone Valley, all the alpine passes played an increasingly important role in trading, along what were originally migration paths. Horse-drawn wheeled carriages would have long since made their way to this region, and people were prospering. The jagged peaks of the Chablais and the Valais Alps (also called the Alpes Poeninae, the name derived from the Latin) provided a breathtakingly beautiful backdrop for daily life on the flood plain, and the low foothills below the Alps. The grape was probably already being cultivated in the well-drained hilly slopes, the wine trade flourished, along with the importation of all kinds of goods transported from and traded with Italy, Greece and Crete, (the Aegean civilizations) as well as Egypt and the near East, en route to France, Germany and Britain. The improvements in transport, travel and long distance communications would have enabled technology transfer from the rest of the known world to arrive very rapidly on one of the main transport hubs of the last few hundred years before our modern era.

It seems logical that on the Upper Rhone plain, increasing territorial claims, political influences, and probable skirmishes amongst the neighbouring tribes would have resulted in tribal boundaries eventually being established. Economic advantage would have been a priority, due to the pressure of limited land availability on the plain, and growing population density. These pressures would have increased during the Iron Age as people developed improved methods of herding and agriculture, quality of food, and corresponding population increase. Another reason for the consolidation of tribal groups must have been the organisation required for new technologies for working metals, and trading goods with other regions. The economy would have required a new communal effort to establish and maintain successful commercial relationships.

Since the end of the glaciation, alpine passes were highly significant for population drift. Gradually overland trade routes would have developed as an adjunct to shipping goods through the Mediterranean, where ports such as Massalia, and Antioch were controlled by Greek merchants. The Ligurian peoples were also seafarers, and they populated the coastal areas of what is now Italy and Southern France. During the Iron Age, the lure of wealth would have induced merchants to explore land routes through Italy and the Alps as well as sea trading routes, to reach the fabled Western Isles of Britain. This must have been a time of great optimism, activity, and advances in the development of systems of knowledge. As well as navigation by sea, people engaged in terrestrial route-finding over hundreds of kilometres where the terrain was wild, and there were no roads, nor signposts. So how did these Iron Age traders and travellers, and their emissaries manage these journeys so successfully?

CELTIC TRAVEL

Whatever the cause for the Celtic wars and Roman imperialism, changes brought by the Iron Age to civilization produced mixed blessings. The advantages of a monumental ability to build in stone was the result of the concrete used as mortar, the longevity of which was due to the chemical composition of the volcanic ash mixed with lime. So it was luck as well as intelligent design that enabled the construction of large buildings, and the ability to transport water along stone aqueducts. The ability to build villas, pipe water and sculpt realistic statuary lauding powerful figures came at a heavy price. Roman culture became centred on the political power of a merchant class spending its money in the construction industry, and this status quo was imposed on all of its European colonies, which in subsequent centuries, in the manner of the conquered, sought to emulate the victors, long after the departure of the Romans. They developed their own hierarchies based on this paradigm, and sadly, emulated the conquerors in the exploitation of slave labour.

Imperialism, conquest, and enslavement by the Roman army at the behest of autocratic emperors may not have been totally responsible for decline in our consciousness of the immediate environment. It may not have been the sole reason for our education systems abandoning knowledge of the stars, the seasons, the knowledge of patterns of weather change, and traditional navigation technologies based on solar and stellar trajectories. However, the very scale of the invasion and conquest had huge consequences on the collective psyche over the past two thousand years, in terms of social structures and aspirations, on all the cultures inherited from early Europe and Britain.

Evidence from surviving roadways in both Britain and France show that just after the time of the Celtic conquest, the Roman army had a curious habit of laying stones on roadways, in a straight line between Celtic oppida and not between the Roman towns and forts, built subsequently. All throughout ancient Gaul and Britain, there were routes that survive today with these straight line trajectories along solar paths, accompanied by short detours from the Celtic endpoints. Strangely, the remaining miles of the routes to the newly established Roman settlements were always at odd angles, as though they were later tributaries to earlier pathways. This begs the question of who surveyed and built the original roads. The phenomenon is so widespread that clearly it was not the Romans. According to discoveries by Graham Robb documented in his seminal work on the lost map of Celtic Europe entitled 'The Ancient Paths', the tribes of the Celtic alliances were the surveyors and the original road builders, in use for centuries by their horse drawn chariots. The Romans, opportunistically, having conquered by force, used the existing routes, paving over the established Celtic roads, using local slave labour, including Celtic engineers.

Robb found that Celtic place names lie in abundance along straight lines angled to the summer and winter solstices of the era, across those regions that eventually became France, Britain, northern Italy and Switzerland. His discoveries of so many Celtic places aligned over hundreds of kilometres in straight line trajectories, means that chance causing these places to be aligned along a trade route accidentally, is in the realm of the fantastically improbable. Scores of European place names with a Celtic etymology, scattered far afield, are traceable to the term 'Mediolanum', including Milan. The term has been identified as a Celtic place name with its roots in proto Indo European language groups. A number of possible meanings have been suggested: What is impressive is that Robb travelled by bicycle to a significant sample of these 'Mediolanum' locations, either confirmed or with a high probability of being derived from the term. He records that these places show little indication of archaeological or historical evidence of permanent habitation in the Iron Age. So the question has been asked, for what were these places established and did the location have a practical purpose? Was there a purpose to siting these locations for distance estimation along the line-of-sight trajectories at staged intervals along a route? Were they multi-purpose places for performing the complex geometry, astronomy and instrumentation, used at the time for terrestrial navigation? Robb's theory is that these places had a definite purpose, and were part of Celtic long distance transport networks, developed over centuries, for the purposes of communication and trade.

What is evident is that the Celts, like every other society possessed the technologies and systems to measure declination of astronomical bodies, as well as the seasonal events, like the solstices and equinoxes, to compare latitude of different places, and to align direction of travel. As well as using rods and string, like the Egyptians, they probably had simple instruments, such as right angles, and some form of geometric square to sight stars, and compare angles of solar and lunar positions. Mediolana were even used for astronomical data, observation of the azimuths of various fixed stars, as well as seasonal locations and variations of the sun, the moon and the planets, using various geometric techniques and measuring devices.

The development of knowledge systems, and the acquisition of technologies from the Greeks and Egyptians by Druids, would have been disseminated amongst the Celtic allies, who had excellent communication networks. It seems very credible that by the late Iron Age, the Celts had established destinations and routes that took them along precise paths through a combination of summer and winter solstice sun alignments. They would have developed the ability to use various triangular geometries applied to measuring places, mountains and other landmarks as reference points. Remarkably, they were able to traverse long distances, and establish routes with accuracy and precision, through wild country. Direction finding along solstice lines seems a practical and clever innovation. The Druids, in charge of knowledge dissemination, were the custodians of the methods of measurement, and calculation, and there were lengthy educational systems for those girls and boys who showed interest and aptitude. Educated local people must have been able to provide routes and directions for their particular local terrain, which, when linked from one community to the next, would have allowed the Celtic peoples to traverse what had to be the first set of long distance public transport routes in Europe.

It is possible to make a comparison with long distance navigation systems of the first Australians, where local people were able to navigate precisely, travelling seasonally over many hundreds of kilometres using a system of recognition, apparent to them from their knowledge of the land, and their finely tuned geolocation instincts. It is clear that the direct finding trajectories known as Song Lines, were not mystical, rather practical, logical and based on maps and the scientific methods of people whose geomagnetic sensory perception was not dulled by agrarian settlement.

The Mediolanum places must have been very useful in an era when there were no roads, signage or regional maps. In fact, the Celtic system of navigation was the GPS of its day, limited only by the weather and the appearance of the sun. To those who could read the landscape and use the system, it was probably just as reliable. The undeniable etymology of European Celtic place names in existence today is the cold hard evidence for Robb's theory of the land navigation achievements, originating in Gaul. Not only were their paths sited precisely on projected directional straight lines, they may also have allowed for the curvature of the Earth. The journey mapping information systems and methods of technology would have been transmitted orally, in accord with the Druidic tradition.

By the time of the late Iron Age, Europe was an extensive geographic concept, though it did not yet exist in name. Intrepid travellers traversed far and wide in search of adventure and opportunity. By this time, the Pillars of Hercules were famed far and wide as legendary rocks that guarded the straits of Gibraltar, the narrow opening of the Mediterranean Sea to the Atlantic Ocean. Many mythic stories surrounded the Greek hero Hercules, son of Zeus, and his feats, particularly the story of his journey on the Heraklean way. Standing near one of the starting points of the Heraklean Way is the ancient Iberian port of Bolonia, a settlement engaged in fishing and a port since the Bronze Age, according to the grave goods found on the site. The town, a take-off point for the sailors of antiquity, for the exotic destination of Tangiers in Morocco, eventually became the Roman town Baelo Claudia. One surprising feature of this town, spread over a couple of hectares, is the amphitheatre, not for gladiatorial combat, but for a company of actors who put on plays for local entertainment. The theatre probably reflected the traditional interests of the local Iberian population, who provided the workforce for the lucrative export fishing industry.

The imposing rock edifices around this location, close to Tarifa, echo the rocky outcrops of the northern coast of Africa clearly visible across the strait, less than 20 kilometres distant, east of Tangiers, the sister port, and along with the rocky outcrop of Gibraltar could well have been part of the inspiration for the name 'Pillars of Hercules' in the original proto historic stories. The rocky projections seem to form a guard for the ships sailing through the passage between the Mediterranean and Atlantic.

The Heraklean Way, said to have been followed by Hannibal of Carthage, (as related by Roman historians Polybius and Livy), may well have started at this port, situated on the Iberian Peninsula, at the southernmost tip of mainland Europe (now Spanish Andalusia). The ancient route proceeded along the coast through Cadiz to Sagres, the sacred promontory, a name assigned by the ancient Greeks and Romans to salient promontories extending into large bodies of water at strategic locations, typically containing a temple. Robb noticed that there is a straight line trajectory from Sagres to the Matrona Pass, the Col de Montgenevre, the lowest and easiest pass through the Alps in northern Italy. This trajectory exactly traverses several locations with Celtic origins on its way. The route runs through the ancient port of Agde, present in 525 BCE, near Marseilles. It also passes through Andorra, where, as the history books note, the Andosini people were defeated by Hannibal in 218 BCE. Historical accounts of the time mention a straight path taken by Hannibal from Iberia to Italy. Perhaps this was one of the early transport routes surveyed and established as a trading route by the Gallic tribes. It ends up so accurately at the pass, following the line of the rising summer solstice sun, that it seems an amazing coincidence. Whatever the circumstances that led to the establishment of navigation routes along the solstice directions, at this time long distance trade and transport was flourishing.

The Celtic cultures, education systems, and knowledge dissemination were oral, and memorised by rote, so there is no written record of the technologies known and understood at the time. In view of the fact that Celts could read and write Greek, they would have had a good grasp on all the known methods of the ancient world. The evidence of the development of straight line navigation paths over long distances means that they had already developed triangulation, and most probably used parallax. The people of proto-history did know how to find east and west, with the rising and the setting of the sun on the equinoxes. They were educated on how to read north and south, using the shortest shadow of the midday sun. They would have been aware that for the ancient Egyptians, the building of temples was accompanied by the use of string to layout a building's geometric alignment, and astronomical positioning by measuring the azimuth of stars. New technology, in ancient times, just as today, has always spread widely and rapidly, and the Bronze Age was no exception.

Human beings had the same desire for new technologies as they have today. The twentieth century assumption that the alignment of temple buildings was a religious practice may well be wrong. As constant knowledge of direction, using the sun, moon and stars was essential for everyday living, and travelling beyond the immediate terrain, at least in part there was probably a practical purpose for the alignment along which buildings, in the Bronze and Iron Ages (and even Neolithic stone circles), were structured. Knowing the direction of sunrise at a particular time of year was essential for survival, for meetings with other groups, for finding sources of food and water. A simple aid would be to have a particular direction set in stone, so that temples were at least multi-functional, providing a year round key to navigation for people who still had seasonal migration paths. And these systems would have easily evolved into intra-regional direction finding as trade routes extended further.

The solstice was particularly important, as it is the time of year when the sun rises in the same place for around a week, giving plenty of time to sight landmarks, and make approximations of distance to be followed up by actual measurements. Abstract measurement for overland routes would have been particularly important, to complement the navigation skills developed by seafarers, accustomed to using the sky for navigation. The Greek ports along the Mediterranean, such as the ancient Greek port of Massalia (Marseilles) founded in the 6th century and Antibes in the 5th century BCE, became important crossroads for sea and land routes. Massalia was established on land bought from or traded with local Celtic tribes. Antibes was originally populated by Ligurian people who had migrated westward.

According to the Roman historians, Hercules had to "cross the mountain that was once Atlas. Instead of climbing the great mountain, Hercules used his superhuman strength to smash through it". Possibly this is the recorded narrative of an oral history handed down over the generations since the end of the last ice age, perhaps to describe a seismic event of grand proportions, such as an earthquake, or the falling of mountains, left unstable by the melting of the ice. Although this makes no apparent sense today, the style of this myth resonates with the indigenous Australian dreamtime stories, which sought to record environmental, climate and geographic events and changes for posterity, using storytelling as the art form for informing each new generation.

Greek and Roman historians did not meet many Celtic people, if any, and relied upon second hand accounts from travellers' tales. Hardly the stuff of scientific evidence, and yet these accounts have been repeated over the millennia as historical truth. What enabled a loose federation of tribes that they became known as a nation? It could not have been language alone. There were several hundred Indo European dialects based on an early form of Sanskrit. By 1,000 BCE there must have been as many differences as there were similar words for everyday objects and concepts. So what else could have unified a number of tribes from diverse regions, environmental conditions, and farming practices, over the past 3,000 years?

The Celts were a group of peoples and cultures, inhabiting much of Europe and Asia Minor in pre-Roman times. The extension of the Celtic territories grew by means of marriage and trade. The culture developed in the late Bronze Age around the upper Danube, with its zenith in the La Tene culture (500 to 100 BCE), extending from Britain in the west, to northern Italy, from southern Germany eastward to the Hycernian Forest, before first being conquered by the Romans and in the wake of the decline of the Roman empire, overrun by various tribes from the north. Over the next two millennia, the original populations retained a strong sense of identity only in parts of France, northern Spain, French speaking Switzerland, and the Celtic strongholds of Britain.

What is really interesting is the realisation that the transport networks developed in the last millennium before our current era (BCE), crystallised over time from nomadic routes as the population re-inhabited terrain left free of ice after the last glaciation. Following paths originally forged by migrating herds, followed by groups of people foraging for food along water courses and coastal paths, a sophisticated system of direction finding and navigation emerged, transforming these ways into trade routes. Just as seafarers learned to sail the Mediterranean, overland way finders were using the sun, the moon and the stars to navigate from place to place, using lunar phases, and solar positioning to establish common times for meetings, through use of angular and linear geometry, and triangulation to establish trajectories. They were the remarkable predecessors of today's journey navigation systems.

It appears that in the upper reaches of the Rhone, as elsewhere, a loose federation of Celtic tribes was established gradually, as an organic popular movement, more or less peacefully. The terrain of the Veragri, the Nantuaten and the Seduni bordered the Rhone, traversing the floodplain, meandering in its course. It appears that in the latter stages of Celtic times, the territory was carefully divided, perhaps after territorial skirmishes, and most probably demarcated by landmark mountain features. No doubt these borders were established over a protracted period. The most recent arrivals were probably the Nantuaten, with strong Gallic ties to the region around Nantua, today a commune or sub-prefecture located between Geneva and Lyon. Successive waves of migration would have pushed the first groups of incomers further along the valley. The Veragri's territory most likely extended from Martigny over the pass at Col de la Forclaz to the Chamonix Valley, and perhaps ended at l'Aiguille de Midi, bordering Allobroges country. The Seduni occupied the terrain somewhere after the valley turns a right angle to the north east to the vicinity of Sion, chef lieu at least since the late Iron Age, and probably much earlier. The main migration paths to the Seduni territory would have been a population drift both from the head of Lake Leman, and over the Alps from northern Italy.

By the late Iron Age, it appears that there was an accord. Cooperation would have been needed to keep the travel routes open through the valley, from the Great Saint Bernard and Simplon Passes. These adjoining territories had carefully located tribal capitals, at strategic locations, and despite the pressures of population, there was probably a history of mutual interaction, given that a major trading route passed through all the tribes' territory. Octodurum, on Veragri land, was a tribal centre, as was Tarnaias for the Nantuaten, near present day Saint Maurice.

It is highly likely that gatherings were held, particularly in summer, to reinforce pacts for maintaining the important business links. No doubt marriage was an important social communication. At first locally, with neighbouring people, then eventually establishing connections across the region, people would have encouraged trade and communication, seeing the benefits of a better standard of living for all. As time went by, the concept of 'safe passage' for travellers, emanating from Gaul, ensured that people of the Celtic groups could travel from region to region, trading materials, artefacts, food and increasingly metals over larger distances. Particularly unifying would have been the annual, cross culture summer solstice gatherings. These celebrations would have kept the lines of communication open, so that travel and the common long-distance transport navigation technologies would continue over time. And like summer festivals all over the world, would have provided the opportunity for ceremony, trade, marriage and exchange of advances in technology.

Just as seafaring peoples were developing navigation skills using astronomical and meteorological data that allowed them to sail over vast distances, and even more amazingly to return, people of inland locations were also developing navigation practices. Never in history have we humans developed unique skills and knowledge, found only in one place, unless we lived in total isolation. We are essentially social beings. We have been communicating knowledge and technologies with surprisingly rapid and efficient networks since, presumably the dawn of our species. Whenever there was contact of any kind, people learned from each other, along trading routes, and even across continents. Celtic art displays an amazing grasp of geometry and mathematical precision that has not been duplicated by any art movement since. It appears that the Celtic culture was diluted and fractured by the troubled times of the late Iron Age, ending with the Roman invasion of Europe.

At the time of the late Iron Age, tribal interactions in the Upper Rhone Valley were not based on domination by authorities and other forms of imposed hegemony, rather on mutual systems of interdependence. Perhaps the predominantly agricultural economy spared them actual conflict. The pressures of population expansion were probably diluted, as people could always occupy land further up the steep mountainous hillsides, despite the fact that it was less desirable real estate. The Rhone plain apparently was spared the hostility generated by the rise of the mining industry, with its insatiable need for labour, to satisfy the demand for metals and implements, one of the root causes of increasing disharmony, eventually leading to infighting and intertribal conflict.

By the time the Romans began their military expansion, the Celts' communication technologies, and knowledge and practice of the sophisticated systems of solstice measurements, and comparative measurements of latitude to navigate by the sun along trade routes, was well developed. It seems odd that this was not enough for the old alliances to mobilise to repulse the Romans. Did it indicate that there may have been a level of disharmony amongst tribes, weakening the bonds amongst allies? Did the increasing presence of warrior's artefacts in the Iron Age archaeological finds mean that inter-tribal aggression was on the rise? Was there a reduction in the cooperation that allowed free travel along solstice aligned routes? These routes would have had to be established over millennia from astronomical knowledge systems and local practices of triangulation from place to place by establishment of equidistant waypoints, and the solar alignment of prominent landmarks. Were the long distance trade routes established by Celtic alliances, with the accuracy and endeavour unparalleled until the 20th century mobile network operations, in decline?

What happened between the zenith of Celtic long distance trade, and the time when the Romans were able to divide and conquer? Did they opportunistically take advantage of division and fracturing of these alliances? Perhaps during the last few hundred years before our current era, pressures of expanding populations and resource constraints may have caused inter-tribal rivalries and warfare, where once there had been a tradition of safe passage and courtesies amongst peoples of different regions. Perhaps this is one of the great lessons of history that we would do well to heed today. It may be time to resurrect the original Celtic vision of advancement by trade, marriage, culture and collaboration, sadly lost at the end of the Iron Age, by their widespread defeat and mass enslavement at the hands of tyrannical Roman rule.

Before the Roman invasion of Europe and Britain, current research (in contrast to the subjective accounts of Roman historians) indicates that the Celtic people were civilised, had extensive technology systems including navigation, astronomy, seasonal land management, trade in art and artefacts, mining and manufacture, as well as lengthy education of their young people and seasonal communal and sacred cultural practices. All of these accomplishments were dismissed by their conquerors as insignificant, and their cultural practices barbaric. That is unless it suited the Romans to 'acquire' and appropriate the Celts' technology. The Latin words for carriages, chariots were from the Gaulish 'karros' originally. The Roman whiteout of the Celtic civilization was the first, but not the last time that empire builders claimed cultural omniscience over the 'natives'. The British Empire springs to mind, indeed giving credence to the old saw, ironically last made famous by Winston Churchill during World War II - 'history is written by the victors'. Perhaps this is true of Roman history, given the fact that for two thousand years the western world has believed and propagated the narratives of the historians who did not leave Rome or Greece.

While in Switzerland, in the alpine regions, directions were often dictated by the surrounding mountains, the line of travel was also carefully calculated. Was this by reference to the solstices, or by mountain landmarks alone? Most probably it was a combination of both. These techniques would have to be adapted for people living in the mountains. While for the locals, directions and distances would have been well-known, for travellers from foreign parts, the mountains must have been extremely confusing. There are a number of prominent landmarks throughout the French Swiss Alps, often incorporating the term 'midi, for example Les Dents de Midi, l'Aiguille du Midi, and lots of Mittaghorns, and there is a convincing case that they were survey points for territory, established by the people of the late Iron Age). It may well be that there were line of sight tribal boundaries associated with these mountains that were further defined by local knowledge.

By the time the Iron Age dawned, and trade had become more intensive, right across Gaul, the Celts had a tradition of not only granting safe passage, but of being on the lookout for danger to travellers. It was clearly to everyone's benefit to keep the trade routes open and as convenient as possible. There has been a plethora of research about the Celts in recent years, based on various historical accounts, archaeological finds, and the etymology of Celtic place names. There is general agreement about the approximate location of the tribal territory in Switzerland in the late Iron Age. The Helveti was the supergroup of the tribes that occupied the Swiss plateau, though not the Valais and Chablais alpine regions. Clearly, over millennia of long distance trading, there must have been strong ties of cultural exchange, trade and marriage amongst the Helveti and the Gallic, northern Italian and Germanic tribes who occupied what is modern day Switzerland.

By the transition from the late Bronze Age to the early Iron Age, the Gallic and Helvetic Celtic tribes must have had a successful, relatively peaceful society. As in the rest of the Celtic world, these alliances would have been encouraged by annual gatherings across regions, encouraged by common interests and information sharing, with established transport networks, particularly emanating from Gaul, an area roughly synonymous with modern France, extending to Britain and Ireland in the west, along the Danube to the East and south to Turkey, through Switzerland to northern Italy. Local and regional cultural affinities, and written and visual communication systems were evidently in place by the start of the Iron Age, for the level of commercial activities discovered from archaeological evidence.

Clearly, most of the Swiss tribes, in common with Gaul,

and Iberia, could make themselves understood amongst tribal groups. At least they all spoke a language that had a common Indo-European root. The tribes of the Upper Rhone Valley probably had the closest linguistic relationships with the Gauls, with many having migrated into Switzerland from southern France. The people who had migrated from northern Italy would have at least some words in common with the Gallic dialects. People have always migrated onwards once local resources become constrained, reaching the limits of the population that can comfortably be supported in a particular locality. Over the past 10,000 years since the population drift into this part of Switzerland, after the end of last glaciation, this probably accounts for the regional division of people into tribes, as the advantages of collaboration and co-habitation in settlements were weighed against the opportunities to gain new resources, and the associated appetite for risk. This distribution of people into new lands would have continued from the time of the Neolithic (stone) age, throughout the Bronze Age, and by the time of the Iron Age, territorial boundaries would have had to be well established, because the Rhone plain is a limited terrain, with limited agricultural opportunities on the lower slopes of the surrounding mountains.

By the time of the Iron Age, the Celtic tribes had negotiated territorial boundaries, and made agreements about trade and resources right across Europe. Territorial borders were often denoted by landmarks and lines of sight, supporting Robb's theory that in Switzerland, borders were delineated by mountains containing the name Midi. This would provide for tribal boundaries of Les Dents du Midi between the Nantuaten and the Veragri, the Aiguille du Midi between the Veragri and the Allobroges outside Chamonix, and possibly the Crete du Midi between the Seduni and the Uberer. The tribes usually had a chef-lieu, an oppidum, central to their culture and society, as well as places to conduct location alignments, providing standard reference directions for north, south, east and west, as well as solar, lunar and stellar observations. The Celtic name for an Iron Age fortress was a 'dunum', and they were often named after Gods. Hence the frequent occurrence of Lugdunum, and its various etymological derivations, such as Lyon, and perhaps Londinium, after 'Lug' the Celtic God of Light.

Anyone coming over the Great Saint Bernard Pass to France had to pass at least through the territory of both the Veragri and the Nantuaten. Having migrated westward from France, the Nantuaten are thought to have occupied territory around Lake Leman at the opposite end from Geneva, to the south of the lake, and westward to the Chablais (from the Latin caput lac, head of the lake). What is known is that the Nantua, the Veragri, and the Seduni occupied the Rhone valley and the surrounding Alps, in what today is the Chablais and Valais regions. The Nantuaten were on the plain where the river joins Lake Leman, and archaeological evidence shows occupation since the Bronze Age of settlements around Aigle, and St Triphon. The Veragri were to be found around Martigny, and their territory probably extended over the passes and gorges that lead to the Chamonix Valley, and the Mont Blanc massif. The Seduni occupied the region where the Rhone plain turns to the north east towards the town of Sion. It is quite possible that the Seduni migrated from the south, from Italy, and the Veragri may have been a combination of population drift along the Rhone from the lake, as well as from over the Great Saint Bernard Pass from the Aosta Valley.

Away from obvious landmarks the mountains can be very confusing, and the possibility of being lost in the wilderness of alpine valleys and peaks must have been a very present concern to Celtic travellers. In the days when there were no sign posts, and the roads over the alps were little more than goat tracks, a landmark sighting system, and established sighting places where routes could be lined up with distant peaks and solar alignments, would have helped people to maintain their bearings, and to avoid wandering off course into the wild alpine country. Some sources believe that viticulture was practised in the late Iron Age around the Rhone plain in Valais, well before the arrival of the Romans.

The tribal capital of the Nantuaten was located at Tarnaias (now Massongex), near the present day Saint Maurice. The earliest record of the name appears on the Antonine route from Milan, as being located between Octodurum (Martigny) and Pennoslacos (Villeneuve). The Antonine Itinerary is a famous route map register of the stations and distances along various roads from the time of the emperor Augustus, a successor of Caesar. The name also appears in the Ravenna Cosmography, the work of an unknown seventh century geographer and a recognized source of the ancient Celtic place names, comprising five manuscripts discovered in Ravenna.

Archaeology and history place the La Tene period, the late Iron Age from around 500 to 100 BCE, as the zenith of Celtic cultural achievement, as attested by the extraordinary beauty and geometric precision of the artefacts found near Lake Neuchatel in Switzerland. It appears that the mobility of the Celtic peoples developed into several main trading corridors, from central France, north west to Britain and Ireland, and south east through the Swiss alpine regions to the Po Valley in northern Italy, and another from Austria-Hungary to eastern Europe, with alpine links to the Adriatic people of the Venetic culture of north-east Italy. The late Iron Age Celtic artefacts are things of delicate beauty, great precision and artistry, in contrast with the later semi-industrial shapes of Roman pottery, implements and statuary. This is in direct contrast to the descriptions of Celtic culture, which were almost all written by the historians of the Roman republic, enslavers of the Celtic tribes. It comes as no surprise that their accounts of the Celts were slighting and dismissive.

While there is no direct evidence that the Ormonts valley was inhabited during the Iron Age, it is only a twenty kilometre hike from Tarnaias (near Saint Maurice), accessible by the relatively easy pass at Col de la Croix. This route may have provided an escape route to this hidden valley that runs parallel to the Rhone plain, on the other side of the Chamossaire Mountain. Towering over the valley are Les Diablerets, high glaciated wild peaks, and the subject of frightening local legends. The local Gallic tribe, the Nantuaten, in whose territory the valley bordered, must have known of the valley, even though it was wilderness, occupied by bears, wolves and lynxes. It seems likely that some of the local legends of the devils and the dangers, particularly from the wild bears, could have been spread by the locals themselves, possibly to discourage exploration in the times when the Roman administration, taxes and regulations were too close for comfort. It seems incredible that Ormonts remained unknown and unexplored for another thousand years, when, with local knowledge it was readily accessible, particularly in the summer months. The earliest settlement in the Ormonts was probably at the little village of Vers L'Eglise (towards the church), and the local hostelry is called 'Auberge de l'Ours' (inn of the bear), perhaps a tongue in cheek reference to local storytelling as much as to past wild inhabitants.

While it is known that the Nantuaten migrated to the Chablais from the region around Lyon, along the south side of the lake, the question has to be asked as to whether they also occupied territory at the head of the lake around Villeneuve, once known as Pennoslacos (head of the lake), and fifteen kilometres further, Viviscus (Vevey), grave site of some wonderful Celtic artefacts? The spectacular finds at Eclepens (Mormont), north of Lausanne, were also geographically quite close, so there was a permanent Celtic population. Perhaps descendants of the indigenous lacustrine population of the Bronze Age remained here, engaged in fishing and whatever crops they could grow in the heavily wooded slopes that towered steeply above the lake. Perhaps the name Vevey offers a clue. It was recorded as Vibiscus in the Antonine Itinerary, by the geographer Ptolemy as Ouikos, Viviscus in the Table of Peutinger, a 14th century copy of a Roman map, and Bibiscon in Ravenne's geography of the 7th century. This provides an interesting view on how names evolve over time. Whatever the origin, it was a settlement on the important trade route linking Gallic and Helvetic Celts. The Vevey grave site comprising wonderful grave goods including metal swords, bracelets, belts and ornaments was found, catalogued, and carefully and sensitively excavated in the late 19th century.

Barely fifty kilometres further, continuing on this trade route, is Mormont where the amazing Celtic finds were discovered in 2006 on the hill overlooking Eclepens and La Sarraz. The archaeological site of Mormont is of exceptional importance in the pre-Roman Celtic world. The treasure is thousands of artefacts that were collected from some 200 graves. The occupation of Mormont, just before 100 BCE, seems very short. What events were able to lead a community belonging to the people of Helvetia to meet at the top of the hill? Perhaps it is possible that the Eclepens treasures were from a number of Celtic tribes, and this site became a communal burial ground at the time of the Roman conquest.

In Switzerland, like the rest of Celtic world, the people forged associations amongst tribes for the purposes of trade, technology, and marriage for mutual benefit.

Indeed, in Celtic society, land passed through the female line, and in the Ormonts Valley men and women inherited equally, at least up to the twentieth century. Over the millennia since the end of the last glaciation, little by little, people would have become familiar with the routes and ways to travel around Europe, including the Swiss plateau, as well as over the high alpine paths. By the late Iron Age, the communications were very sophisticated. Messages were transported not only on foot, and by horse, but also by vocal signalling. In Switzerland amongst the people in the Alps, yodelling across mountains was a unique innovation, and it was a vital source of information. In the high country, fires strategically placed could also reach great distances, and were also used to relay messages from mountain to mountain. Standing on a high peak on August 1st, the Swiss National Day, one can still see fires burning the length and breadth of the Alps. Solstice measurements, knowledge of meridians and latitude by virtue of being able to identify north, south, east and west was part of the Celtic education system. In Switzerland, adaptations to methods practised in Gaul would have been developed to deal with the alpine geography, and to take advantage of mountains as landmarks.

For a few hundred years, the Celtic world was at its zenith, occupying as it did, most of south-western and central Europe, with organised tribal boundaries, systems of measurement that were shared across the loose confederation of Celtic people, a time of relative peace and prosperity. Clearly the Romans were watching the Celtic expansion with envious eyes. They were plotting, scheming and planning their own ambitious world expansion, which sadly became much more ruthlessly militaristic than the Celts' combination of fighting capacity and diplomacy.

Throughout the Celtic world, there was a 'telecommunications' network that allowed people to pass significant messages. Robb's work 'The Ancient Paths' describes a network of vocal messaging right across the countryside, and by the time the Romans invaded, messages could be relayed hundreds of kilometres in a couple of days, by using the acoustic characteristics of the landscape, as Robb points out, a speed that would not be achieved again until the 18th century. Fast messaging would also have been a characteristic of inter-tribal communications in Switzerland. Good communications would have been essential for the Nantuaten, Veragri and Seduni to collaborate in their successful bid to harass the Roman army into withdrawal in 57 BCE. Because of a strategic knowledge of local geography, and established communications, they must have been able to mobilise inter-tribal forces on a reasonable scale. Men, women and children would have assisted this effort, and though Roman historical accounts tell of a force of 30,000 men, it is likely there were far fewer. With the huge strategic advantage of knowing the local terrain, the hidden paths up into the surrounding mountains, and the best places to ambush the Romans as they poured through from Simplon and the Great Saint Bernard passes, the Romans could have been repulsed with far fewer. The local people would have seen them coming many hours in advance, enabling them to organise strategic skirmishes from their high vantage points. Though Caesar claimed the campaign a victory, and the Roman army burnt Octodurum (Martigny) as they retreated, they were discouraged from any subsequent attempts for another forty or so years.

As far as the Romans were concerned, the Alpes Poeninae was both a barrier and an opportunity to cross the mountains. Roman historian Livy cites the passage south of the Celtic tribes of the Boii and Lingons passing through the Alps in the early fourth century BCE. This was the time when the Celts were seeking new territory, and was part of the organised mass migration that saw Celtic tribes established in northern Italy. The Romans took the territory back from the Salassi in 191 BCE. Mass migrations took place in many directions, and it is not known exactly why they occurred, however it was no doubt a search for new territory, as old terrain reached the limits of supporting current populations. Skills in terrestrial navigation, extensive local survey lines, solstice lines, and east west meridians, and measuring places for calculations using triangulation and astronomy must have emboldened the Celts, and with established routes, technology and knowledge of terrestrial navigation, they could move to seek more opportunity. Northern Italy was seen as a prime destination, in part because of its suitability for growing wine. While most of the Gallic Celtic tribes migrated along the southern route, over the Matrona Pass, the Boii and Lingons used the Great Saint Bernard Pass.

In the late Iron Age there were at least six major tribal groups inhabiting what today is modern Switzerland. Oppida were scattered throughout their territories, from Geneva, to Zurich, Lichtenstein to Valais. The Roman conquest of Europe was the start of the demise of the Oppida. Celtic scholars have pieced together the evidence for Helvetic Celts having made large scale, carefully planned migrations on a number of occasions in the late Iron Age which are recorded in Greek and Roman histories. Perhaps there were other unrecorded well-organised migrations. In any case, the last was in 58 BCE, as the Helvetic tribes attempted to move to southern France. With Swiss efficiency, they had lists, cited by Roman historians, covering all of the men, women and children, and no doubt food, goods, chattels and equipment as well. Clearly by the late Iron Age, European Celtic societies were well organised, planning and executing missions with great precision, exactitude and ambition. This attempt was halted by the Roman army, on the orders of Julius Caesar, near a Celtic town named Bibracte. The end of an era.

THE ROMANS

Caesar's ambition to control the Great Saint Bernard Pass and the trade route across the Alps failed in 57 BC due to strong opposition from the local Celtic tribes, Veragri, supported by their neighbours, the Nantuaten and the Seduni. The Roman army burnt Octodurum, today called Martigny, as they retreated. It was not until over forty years later, that the mad, ruthless Augustus succeeded in conquest of this trade route. Aosta and the Salassi people, on the Italian side of the Alps, were subjugated in 25 BCE. Sometime around 15 BCE, the people of the Upper Rhone Valley came under Roman rule. They were the Seduni at the Sion end of the valley, the Veragri who inhabited the territory around Martigny, and probably over the La Forclaz pass to Argentiere, if not Chamonix Mont Blanc, and the Nantuaten whose territory included Tarnaias, under the shadow of Les Dents du Midi, and the terrain on the southern side of Lake Leman.

There is no doubt that the geography of the Upper Rhone Valley, Valais and the Chablais, was the reason that the people managed to hold out against the might of the army of Rome. How they must have delighted at luring the Roman soldiers into an ambush at the Gorge du Trient, where there was no escape for an army who followed them. They would have been trapped between two narrow towering rock escarpments, rocky walls soaring upwards for hundreds of metres that continued along a stream bed for kilometres. The only escape route was upward, and only the locals would know where and how to scramble up the otherwise inaccessible cliffs. The geography of the mountains would have enabled a relatively small number of men, women and children to resist the might of well-armed, well organised soldiers.

During the years before the Great Saint Bernard Pass belonged to Rome, the local people would have heard the news echoing all over Celtic Europe, that a brutal invading army showed no mercy. With the threat of invasion, they would have recognized that it was only a matter of time, and it was no doubt at this period that the local people found paths that climbed into the Alps, building settlements in places as high, remote, hidden and defensible as they could find. After the first attempts by the Romans to control the trade route over the Alps at the Great Saint Bernard Pass, the locals would have begun to fortify their settlement on the sloping plateau on the side of the foothills surrounding Les Dents du Midi, under the towering peaks of this massif. The Celtic word for fortress is 'dun' or 'dunum'. Taranais Dunum, the fortified Oppidum of the Nantuaten, was strategically sited to be defensible from any invasion by the Roman army. After the first unsuccessful attempt at conquest, the focus of the local people would have turned away from the traditional occupations of herding and farming, long distance trading and controlling the passage of goods and services through the Upper Rhone Plain. The objective became survival under the onslaught of the biggest invasion Europe has ever known. Huddled together for mutual protection and defence, people started to climb higher into the surrounding mountains, perhaps making their mazots and mayens (summer huts for herding) habitable in winter, and started to develop the alpine farming methods that lasted till the 20th century.

The spread of ideas and knowledge slowed to a standstill in Celtic Europe in the wake of large scale conquest by the Romans, as men, women and children were either killed, enslaved, or had their movements constrained by the army, administration, and taxes. Thus began the era of mass enslavement on the grandest geographic scale ever known to history. It was followed by the dark ages, domination by the Church and feudalism, and the rise of a princely class, wealthy individuals who ruled over others by claiming a divine right to power after the exercise of force. This may not have been the first time we were subjected to domination by a wealthy status quo, however it was the first time in recorded human history that social organisation became hierarchical through use of fort main, the power of the sword, on such a scale.

However, the Nantuaten and the Veragri, with the Seduni did manage the incredible feat in 57 BCE of repulsing the well-organised, funded and provisioned Roman army, who eventually retreated. It is probable that the tribes on the Swiss side of the Great Saint Bernard Pass, ran a concerted campaign against the Romans with the neighbouring Salassi, located on the Italian side of the pass. Caesar actually documented the retreat as a victory, however he was never willing to put a time frame on any return. In 'De Bello Gallico III' Caesar recounts that the Roman army took the fortresses, but ambassadors of the Nantuaten and the Veragri negotiated an exchange of hostages in return for a peace deal. Caesar went onto record that he agreed to leave two cohorts with the Nantuaten, and one with the Veragri at Octodurum. In retrospect, it seems that this may have been a euphemism for having lost three cohorts to the local Celts, because no further attempt was made to use the Great Saint Bernard Pass for more than forty years. The Salassi remained independent till 25 BCE, and the Nantuaten and Veragri till 15 BCE, when they finally had to succumb to the deranged and megalomaniac emperor Augustus.

How did the local people manage this feat? The answer had to lie in grit, determination, courage and the excellent good fortune of being able to climb upwards into the alpine terrain, of which they had extensive and critical local knowledge. It seems likely that the men, women and children used their local knowledge to defend the alpine pass from Aosta, by ambushing the Roman army from above, and perhaps by tempting them into difficult and dangerous steep slopes, on goat tracks that led nowhere, where, burdened with heavy shields, the men would have struggled to make any headway. These tactics would have been equally successful on the steep, thickly wooded terrain near the plain, or higher up, on the bare rocky slopes with treacherous ascents and descents. The realisation must have dawned amongst the local Celtic people long before, that the Alps were their best friends, and were always going to be a place of safety, from which to wage strategic campaigns, when invaders came.

The local people had natural castles that could be defended. This advantage is no doubt the reason that Rome, when the army was finally successful in taking Octodurum and Taranais, gave so much autonomy to the Celts of Valais and the Chablais. A legion could have been expended trying to take just one of the mountain settlements, so a diplomatic resolution would have been for Rome to deal with Celtic leaders on civic matters, such as taxes and administration. The local people avoided mass enslavement, a fate that befell fellow Celts in northern Italy and Gaul. The memory of the earlier defeat at the hands of the local tribes was still in the Roman psyche, and for the average Roman soldier, being attacked from above as they followed the locals into the mountains, or along gorges would have been a very unpleasant experience. Every movement of the Romans on the plain below would have been known to the Celts, clearly visible. The army's strategies could have been read like a book with the eagle's viewpoints from the mountains towering on either side of the Rhone plain.

Today what remains of the Celtic age in the upper Rhone valley are the foundations of the Gallic temple at the heart of Martigny, discovered during the building of the Pierre Giannada Foundation museum in the 1970s. Built around 51 BCE, shortly after the initial repulse of the Romans, the temple comprised a podium built over dry stone wall foundations. The superstructure comprised wood and colombage (timber frame), and the inner sanctum had an entrance facing to the North East. The alignment of the temple, like most Celtic temples, is probably significant in terms of relation to direction finding with the technologies of the day.

Octodurum was the earlier name for Martigny. Several hundred metres above the town, on the pass that leads to the Chamonix Valley, is a place called Meylan, which like Milan is considered by Celtic scholars to have an etymology originating from the term 'Mediolanum'. The purpose of this place may well have been originally for location finding and distance sighting for terrestrial navigation for the long distance trade routes that passed along the Upper Rhone Valley.

The normal rhythm of activities for astronomical and planetary observations probably ground to a halt after the initial attempt by the Romans at conquest. Meylan is a wonderful vantage point from which an advancing army could be seen coming from a number of directions, so was doubtless used for vigilance. The local people's attention would have been firmly focused by the experience of skirmishes and battle to turn their minds to strategic defence.

Because of the date of the temple, built a few years after the repulsion of the Roman army, it is likely that it may have served more than one purpose, and the alignment of the temple may have had strategic significance. The excavated temple wall was rebuilt in the seventies, and the original alignment may have been skewed, given that subsequently Roman buildings were erected over the temple's foundations. The alignment of the temple may have been directed to the summer solstice, or possibly point to a geographic location such as the Gemmi Pass, a quick mountainous short-cut route to Interlaken. The preoccupation of this time would have been to make provisions for a quick escape from any further incursions by the Roman army. Clearly migration would have been on their minds, although it turned out that the Roman conquest was so widespread that the best place to escape would have been upwards into the wild mountains. Just how widespread was probably a realisation that slowly dawned on the local people of the Upper Rhone Valley, as news and messages came through from Gaul, the Helvetic plateau, and the Celts from St Gallen. On the plain, the people of the Chablais and Valais would have engaged in fortifying the existing Oppida in the foothills above the Rhone plain. At the border of the territory of the Nantuaten and Veragri tribes, was a centre for trade and commerce, Taranais, after the Celtic God of thunder. Existing in the Bronze Age, on an alluvial flood plain, reclaimed by climate change from what was once a blue green river of ice that gave birth to the River Rhone. It was perhaps named for the thunderstorms that roll around the mountains, descending from the Argentiere glaciers and Mont Blanc towards the Rhone plain. As Octodurum had been burnt by the Romans as they retreated to northern Italy, the Veragri would also have been busy rebuilding their fortress at this strategic crossroads location.

Causing great confusion to historians, was a passage written by Caesar, citing the Rhenus (the Rhine River) 'rising in the country of the Lepontii who occupy the Alps, and that it flows by a long distance through the country of the Nantuaten, Helvetii, and others'. It seems that geography was not Caesar's strong point, as clearly, just like school children today, he confused the Rhine (Gaulish Renos, literally "that which flows," from the proto Indo-European root 'reie', to move, flow, run) with the Rhone River. The Romans finally secured the Upper Rhone Valley, however it appears that the locals had prepared their strategic positions by fortification and settling higher into the Alps, so that their negotiation position was much stronger than the tribes who fell in battle. It seems that the tribes of the Upper Rhone Valley managed to deal with the Romans on an administrative basis, and were smart enough to press their earlier advantage, backed up by impregnable hill forts, at least at Taranais Dunum, to gain a significant advantage for the people, who maintained local systems. This was probably achieved by locals becoming tax gatherers for the conquerors, and guaranteeing safe passage for Romans of the lucrative trade route from Italy through to Germany.

As far as the Romans were concerned, the Alpes Poeninae

became a small Alpine province of the Roman Empire, one of three such provinces in the western Alps between Italy and Gaul. It comprised the Val d'Aosta region (Italy) and the Canton Valais (Switzerland). According to Livy, it included the territory of four tribes, who were not Celtic Helvetian. In any case the location of the Seduni around Sion, and the Veragri around Martigny, and the Nantuaten located in the Chablais is not in dispute. The Nantuaten were a Gallic people whose territory extended from France, and migrated towards Geneva. Today there is a village called Nantua around 70 kilometres from Geneva. As the route along the Rhone from the Mediterranean was one of the earliest pathways to Switzerland, no doubt that many of the people in Valais migrated from France at some point in their history, meeting the Germanic Celts from the West, and the Ligurian Celts from Italy in the Upper Rhone Valley. The route from the Mediterranean to Lugdunum (modern Lyon) along the Rhone to Geneva and around the shores of Lake Leman, has been a major migration route since the end of the last glaciation. These tribes would have been closely connected by both trade and marriage, and the DNA of all Celts would have been mixed amongst the tribal populations.

The Romans had clearly coveted the trade routes through Switzerland for some time, and probably the tales of the wealth that passed through the Alps were legendary in the late Iron Age. The Salassi were a Celto-Ligurian tribe whose lands lay on the Italian side from the Little St Bernard Pass, across the Graian Alps, and the Great Saint Bernard Pass around Aosta. They were finally defeated and many enslaved in 25 BCE by the Romans, who had earlier taken their gold mines in 143 BCE. No doubt this territory was a prize for Rome, and enabled them to begin the take over the Upper Rhone Valley. After their success in taking the territory of the Veragri and the Nantuaten in 15 BCE, the former Celtic territories were now recorded as a Latin province of the Alpes Poeninae, with 'the procurator of Augustus with the right of the sword'.

Thus ended an epoch in the Upper Rhone Valley. Conquered by Tiberius and Drusus, a small confederation of four 'civitates' was established in the Upper Rhone Valley. The confederation received Latin rights in the reign of Claudius. The troops stationed there were not Roman legions but contingents of allied forces whose main role was to guard the mountain passes to Italy. The conferring of Latin rights gave protection under Roman law, so a negotiated settlement took place in 15 BCE rather than an outright battle, as the locals appear to have retained self determination to a certain extent.

Octodurum (now Martigny), the chef-lieu of the Veragri, became the centre for the Forum Claudii Vallensium. It was at this time that the four cities were combined into one, the civitas Vallensis, with a bureaucracy governed by Rome. The major administrative centres were Forum Claudii Vallensium (at Martigny), Pennoslacos (at Villeneuve), Seduni (at Sion), and Tarnaias (at Massongex close to Saint Maurice).

After the establishment of Roman rule, Celtic knowledge of trade and travel routes based on the alignments and trajectories that were calculated and measured over the preceding millennia all but disappeared. The skills may already have been in decline, as the Celtic empire centred on Gaul had been spending increasing amounts of energy on either acquiring or defending territory from incursions by incomers, or skirmishes instigated by tribes wanting to migrate to new territory. In the wake of conquest, the attention of the local people would have turned away from weaponry and defensive systems, and people would have turned to building, farming and forestry. Different skills were developed as permanent alpine settlements would have allowed farmers to turn their hand to developing new practices to deal with the high alpine country, new implements, tools and measuring systems, which were subsequently handed down over the generations.

The coming of the Roman Empire to Octodurum and Tarnaias would have completely changed the lifestyle of the people. There are always winners and losers in business, and local industries, such as metal working and manufacture of horse drawn chariots and agricultural implements though continuing much as usual, probably changed hands in accord with the new status quo. Celtic traditions developed over thousands of years, the sharing of long distance trade and transport routes, the established order of metal technologies, cultural practice, trading, astronomy and navigation, traditional arrangements for marriage and liaisons amongst families would have gradually disappeared, giving way to Roman law, religious belief, culture and technology.

On the Rhone plain, religious practice began to have a stronger focus, as a response to adaptation, and the local people adopted new Roman gods and goddesses, cults and customs. Adopting the cultural and religious beliefs of the conquerors is an evergreen and significant survival tactic adopted by the vanquished. The stress of losing cultural identity is significant, and heightened religious practice is also symptomatic of a pattern where there has been a requirement for engagement in conflict over a protracted period, and a loss of traditional culture. This is usually accompanied by a compensatory mechanism that spurns traditional knowledge and culture, replaced by cultural identification with the victors, an increasing religious fervour, and heightened ambition to status (all natural human reactions and responses to loss of tradition and culture).

Of course many people just adapted to the lifestyle under the conquerors. For those who did not, could not or would not, new settlements would have been established on increasingly higher ground. When trouble with the new regime threatened, or limits of tolerance were reached, people would have left the settlements on the plain, to occupy the fantastically inaccessible places, in sequestered spots, hidden behind rocky outcrops, out of reach, developing skills and technologies as mountain farmers. Most probably, over the generations they gradually forgot part of their former cultural identity and technologies. Part of the forgetting was the knowledge of long distance travel based on seasonal observations, astronomical knowledge and methods of using common measuring points to find trajectories along an agreed standard, the directions of the summer and winter solstices to travel across Europe.

There are plenty of indications that the Celts were not understood, and in fact misrepresented by Roman and Greek historians, who had never been to Gaul, and recorded exaggerated travellers' tales as fact. They were also misled by the Celts' love of double entendres, and deliberate obfuscation of meaning. Many people probably adapted to Roman rule for a few centuries in a typical Gallic pragmatic way. They used their intelligence, knowledge and technologies, as administrators, traders and mountain herders and farmers, although there is little information about what kind of taxes and burdens were imposed on them. The Roman emperors who were more or less megalomaniac, and as they assumed a messianic stature, the plotting and intrigue grew while the empire extended from one end to the other of western Europe. It was an unrealistic overreach, and like all the subsequent empires of history where conquest was by force, power struggles ensued, and eventually the empire declined and fell at the hands of the Germanic peoples, who had ironically supported them in their original conquest of Gaul.

For some people in the Upper Rhone Valley, submission to the conquerors would not have been acceptable. Remote strongholds and fortresses in the surrounding mountains would have begun to be built and occupied by the local people who had the resources, the local knowledge of alpine tracks and terrain, and the desire not to live under Roman subjugation. This period in Swiss history was probably the birth of nearly two thousand years of independent alpine pastoralists and foresters, who harvested the huge pine trees for the chalets with hand forged axes, cleared wherever the slopes were less then impossibly vertical, and built settlements of families huddled together for mutual protection and defence from the rigours of mountain farming. As the old ways of living on the plain and the surrounding foothills disappeared, the most independent minded people developed an existence out of the mountain terrain. For those who found collaboration with the conquerors distasteful, denied the fertile plainland, around the banks of the upper reaches of the Rhone, they did what people have always done as a response to territorial incursion. They moved on and became the ancestors of today's mountain farmers, people who developed methods and technologies, to farm the difficult terrain, land just below where the tree line ends, taking their sheep, goats and cows in the summer, to the high country, where the sweet alpine grasses provide rich grazing.

Roman civilization began to retreat from Swiss territory when it became a border region again after the 'Crisis of the Third Century', a period in which the Roman Empire nearly collapsed under the combined pressures of invasion, civil war, plague, and economic depression. By the end of the third century, during the reign of the dual emperors Diocletia and Maximian, life became more onerous for the conquered. It seems that the people of the Upper Rhone Valley joined the 'Bagaudi' protests of the Gauls against harsh and unjust rule, prompting moves away from the existing established settlements, administered and taxed by Rome. The obvious place to go was up into the surrounding mountains. Particularly between the Rhone and the Chamonix valleys there is a labyrinth of gorges, and steep paths up to hidden rocks, peaks and dips that would have made bolt holes for those who fell afoul of the administrators. Starting with a semi sheer escarpment leading up from the Rhone, the terrain above the plain is only occasionally less steep than 60 degrees, allowing for a foothold for settlement. Today every niche has been built on, every slope that is less than impossibly steep, has pasture, or is rock terraced to allow the building of a chalet into the slope. Living in these parts would have required athleticism to traverse the mountain tracks in summer, tracks which became even more difficult from Autumn to Spring and impassable in Winter. Settlements were strategically located up semi impassable tracks. There were also more remote places in valleys that were offshoots of the Upper Rhone Valley. The henchmen of the regime would have been very lonely and vulnerable indeed trying to climb to the habitations situated on natural rock fortresses, as they could have been easily ambushed and attacked from vantage points, and meet unfortunate ends.

Given their strategic skills, the local people would have continued to self-organise, and pioneer wilder places, gathering together, building mountain chalets for mutual defence, clearing any terrain that could be prized out of the mountains, establishing farming practices, building the mazots and mayens, the mountain huts for guarding livestock, and styling the buildings to allow the cattle, sheep, and goats to be ensconced in stables on the ground floor of the chalet to survive the long cold snowy winters above the plain below, practices that continued over two thousand years to the present day.

Perhaps at last here was a parable where we can learn one of the great lessons of history, that conquest and expansion by force is a dead end, although the events of the time, for the last two thousand years show that history is a work of fiction written by the victors. Even today, the reference works quote the sycophantic accounts of the emperors' murderous exploits, and educational works teach of the might and power of the Romans as though they were glorious purveyors of superior culture. Histories still repeat the Roman litany that barbarian Celts were civilised by being subjugated and enslaved. Patently untrue, and it may be high time to take a more objective view. Perhaps the historical glorification of war may go some way to explaining our obsession with imperialism, and a capitalism based on the value of metals at the expense of sustainable farming and the natural ecosystems, as well as the paradigm of harnessing the labour of the many for the wealth of the few, to the detriment of the many.

Meanwhile the Swiss were consolidating their new occupations as mountain farmers. The higher and more inaccessible their habitations, the less influence the feudal establishments had on their lifestyles. For the mountain farmers and foresters, life was conducted in a constant battle with the elements. These people gradually ascended the hillsides above the valley, clearing tracts of forest as they went, to provide grazing for their animals, and gardens for their fruit and vegetables. The local feudal masters continued to claim the divine right of kings, collect taxes, and generally take the profits of the labours of the local people. People who disliked a state of subservience moved higher into the mountains, where they were less accessible to the lackeys of the state. After all, the tracks into the mountains were still dangerously steep, the valleys still inhabited by bears, lynxes and wolves, and most of all, winter saw the montagnards cut off for six months at a time from the Rhone plain, and the new lords of this historic region, the bishops of Sion. Every indication is that our Swiss family, like all those of an independent disposition, saw out this turbulent period of human history forging new communities and ways of life as far from the seat of feudal power as they could manage. They were perched in mountain valleys during the summer, and ascended with their herds to the high country, establishing alpine pastures from the mountain wilderness that surrounded the Rhone plain. By this time, there would have been successive waves of incomers, preferring the hardy lifestyle of the mountains to the conditions of the dispossessed on the Upper Rhone plain. During this epoch, the control of the ruling families became increasingly autocratic and imperial, and free movement throughout the former territories of the old Celtic allies became increasingly difficult and then downright impossible. Eventually the Burgundian kings formed alliances with the Holy Roman Emperors, equally autocratic and despotic military rulers, and Europe continued under the yoke of a power mongering warrior class, who today would be diagnosed as completely psychotic and probably insane.

Earlier freedoms and the liberties of trade and marriage became no more than a distant imprint in the DNA. The people of Celtic descent had no record of their former journeying, through the Matrona Pass, the Col de Montgenevre, the end of the Heraklean Way, the ancient route from the Iberian Peninsula, the route along the Rhone Valley from Lyon to Geneva and on to the Alps, the trade route through Eclepens and Le Mormont to the French coast, nor any other of the Celtic routes that followed the angles of the rising and setting solstice sun. Earth had lost another set of technologies, culture, art, knowledge and civilization to the dictates of war and conquest. Another setback for intelligent technology innovation.

THE EMPIRE FALLS

Agaunum is now called Saint Maurice after the legendary Theban legion commander who, according to the Catholic Church, refused an order to kill local people who by this time had converted to Christianity. Mauritius and his men, being Coptic Christians would have no part of a massacre of local people on the grounds of religion. The result, according to an account over one hundred years later by Eucherius, was that emperor Maximian ordered the legion to be decimated, twice, eventually killing all 6,600 men. Near Agaunum, a place still identifiable as a former temple to Mercury, god of travellers, was excavated behind the abbey's present sanctuary. The site was the recorded discovery of martyrs' bones during the time of Theodore, Bishop of Octodurum, who was in office in the year 350. This was probably an authentic discovery of remains, and there is now no way of knowing whose remains they were. In all likelihood, particularly as there is a local name for the place where the skeletal remains were found, there is at least some truth to the legend of Saint Maurice. Another account has them as: 'Tertia Diocletiana Thebaeorum, the third Theban legion of Diocletia, the soldiers of this legion are from Egypt, allocated from around the ancient city of Thebes, but of Christian religion, who arrived in the country of the Veragri' was written by Pietro Gioffredo, an historian from Nice, in 1671, after a visit to the Abbey at Saint Maurice. While most accounts originate with Eucherius, the local name for the place where the bones were discovered is Verroliez from 'vrai lieu' or true place. There does not seem to be any other written account of the events, however there is no doubt it is based on circumstances that left a deep imprint on local consciousness.

At the zenith of the dynastic Egyptian civilization of antiquity, the archaeological remains of Thebes show it to have been a shining jewel, an architectural, social and cultural light of the ancient world, with sophisticated astronomy, construction and civic systems. During the Bronze Age, great residences, brightly painted and surrounded with gardens, were built on the banks of the river Nile. This was the height of palace culture, envied and emulated throughout the Fertile Plain of Asia Minor. Many wealthy families had large estates, and in the gracious public spaces, foreign traders and mercenaries mingled with the citizens. The Thebans at this time were prosperous, well recompensed for their labour in civic projects, such as building temples and tombs, public gardens and irrigation works. As for every other civilization, the decline and fall of the age of the pharaohs was inevitable. Incursions by people from the north, who envied Egypt's climate stability and the annual flooding of the Nile destabilised Thebes. The Assyrians sacked Thebes in the 7th century BCE. Although it was later partly restored, the city declined steadily after the collapse of the 31st Dynasty. Thebes was sacked by the

Romans late in the 1st century BCE.

According to tradition, Christianity was introduced to the Egyptians in the first century AD. From Alexandria, Christianity spread throughout Egypt. It spread to the rural areas, and scriptures, written in Coptic, were translated into the local language, today known as the Coptic language. By the beginning of the 3rd century AD, Christians constituted the majority of Egypt's population, and the Church of Alexandria was recognized as one of Christendom's four Apostolic Sees. The Coptic Encyclopedia finds that there were two legions bearing the name 'Theban', both of them formed by Diocletian sometime after the Roman conquest. Diocletian rose through the ranks of the military to become cavalry commander, and successfully plotted to become an emperor. He was responsible for raising legions of men for the army as part of the settlement at Alexandria. Rome often employed soldiers from the provinces, sparing Romans from engaging in the more difficult and dangerous work. Men were co-opted into the Roman army from Thebes. Current Coptic tradition has it that there was a single cohort, a tenth of a legion, martyred at Agaunum. The remainder of the cohorts (battalion sized units of which there were ten to a legion) were either on the march or already stationed along the Roman road that ran from Liguria through Turin and Milan, across the Alps and down the Rhine to Colonia Agrippinensis (Cologne).

Historical record indicates that in 286 Maximian

recruited a legion of men from Thebes because of a Gallic revolt by people known as the Bagaudi. Originally the term meant a group or a troop, and in Valais, would have comprised the local Celtic people who were unhappy with Roman rule. They probably hid out in the less accessible parts of the surrounding Alps, rather than submit to Rome. The Thebans were recorded as having been stationed in north Italy just over the Great Saint Bernard Pass, and it is likely that they were dispatched the short distance over the Alps to Octodurum and Agaunum (Saint Maurice) to quell the troublesome locals, to deal with the upstarts seen as setting bad examples to the rest of the population. The locals were no doubt safely ensconced in their well defended, largely inaccessible mountain fortress hideouts, high above the Rhone Plain. By the third century, the legion was a much smaller unit of about 1,000 to 1,500 men, and there were more of them. It is more likely that the Thebans were a body of far fewer than 1,000 men, dispatched to deal with the Bagaudi around Agaunum and Octodurum, assembled ad hoc to meet a crisis situation in the Upper Rhone Valley. The monastic accounts themselves do not specifically state that all the soldiers were collectively executed. An eleventh-century monk named Otto of Freising wrote that most of the legionaries escaped, and only some were executed.

The Thebans were not barbarians, and although mercenaries in the employ of Rome, the practice of employing Thebans as auxiliaries was originally part of the Egyptian surrender, and clearly continued as an occupation for young men from a state with a troubled economy. They were ordinary men, members of a civilised society, who according to historical account, were Coptic Christians, being ordered to kill other men, women and children, something that was totally against their beliefs. How do you get people to kill other people? That is part of the conditioning of the regular army. Yet the Thebans would have had no love for the emperor, feeling no allegiance to Rome. This was clearly a crisis of conscience, Roman army or no Roman army.

The geographic features of the Alps meant that local knowledge would have allowed the Bagaudi to pursue their own destiny, safe from the Romans. There are difficult steep tracks all over the mountains in this region, known only to mountain goats and locals. The knowledge of these pathways would have been a saving grace for the local people, and it is no surprise, after Rome's previous experience once the locals were dug into their strongholds, that a foreign source of troops was sought to once again rout the Celts from their traditional, strategic, natural citadels. Over the two hundred years between the coming of the Romans, and Maximian's deployment of the Theban legion to maintain the trade route over the Alps, the local Celtic people had plenty of time to develop secret tracks through the highest, steepest and most inhospitable parts of the mountains and surrounding Alps. These trails today are used by recreational hikers, well sign-posted and maintained by local authorities. In those days, the tracks would have been a complete mystery to the Romans, and because of the terrain, exploring them without local knowledge could have led to an impassable precipice, a cul-de-sac gorge, and hence into an ambush at any time. Tracks that would have allowed the Veragri, Nantuaten and Seduni who did not live on the river plain, to resist Roman rule indefinitely, using local knowledge and advantage to ambush and trap the soldiers from Civitatis Rome. No wonder the emperor wanted to use foreign troops, when trying to face down the alpinist Celtic people on their home territory. It appears that the lighter hand of government exercised by previous Roman rulers, whereby the local people had a large degree of autonomy over their own affairs, came to an end with the bloody rise of Diocletian, a man of peasant stock who took the title of emperor through aggression, treachery, and assassination.

The story of the martyrdom of the Theban legion appears to have deep local roots. According to a letter from Eucherius, bishop of Lyon written about 450, bodies identified as the martyrs of Agaunum were discovered and identified by Theodore, the first historically identified Bishop of Octodurum, who was present at the Council of Aquileia in 381. There is a lot of scepticism about this legend from various sources, and it is highly likely that the account written by Eucherius had the usual distortions of a tale told after the fact. And perhaps it was an exaggeration. That is not to say, however, that it was without foundation. There may be a number of facets to the narrative explaining the cache of human remains that led to the sanctification of Mauritius and his men, who most probably did exist.

The Thebans were being asked to do what the Romans did not want to do, take on the Celts on their own ground. The Thebans, as mercenaries with no allegiance to Rome, could well have decided that to refuse to slaughter the local Bagaudi of Celtic origin. The Thebans homeland, too, had been invaded by Rome. The Thebans were poorly paid mercenaries who had not been brainwashed by extensive military training, neither to owe allegiance to their conquerors, nor to regard the 'Bagaudi' as enemies. If the Thebans refused to kill them, and some were put to death as a consequence, the local people could well have used emissaries to offer the Thebans guidance and shelter, and escape routes to various safe places, including a hidden valley where they would not be found, when they chose to desert rather than face the despotic emperor's wrath, and further executions. The account, deposited in the archives of Lausanne, written by Henri Tardent, a Swiss professor living and working in Nikolayev in Russia in 1887 contains an anecdote handed down in the Ormonts Valley. He records a story that was said to be as old as the families who were the earliest inhabitants of the Ormonts Valley, present at least since 1398. Henri's account asserts that the valley, once an uninhabited wilderness, was always said to have been founded by men who had run away from a Theban legion of the Roman army stationed at Saint Maurice.

If they had made contact with the local Bagaudi, then desertion would have become a viable option. Both the people of Thebes, and the original inhabitants of the

Upper Rhone Valley had no love for Rome, particularly with the tyrants on the emperor's throne at the time. It is very possible that the locals offered shelter in the remotest location known in the region that was habitable at the time. In which case, it really begs the question of exactly when the Ormonts Valley was originally settled. Perhaps at least some of the ancestors originally found their way to this challenging environment earlier than the twelfth century, as the history books have it. The local Celtic people perhaps had already established a couple of settlements in the hidden valley that they offered as sanctuary to the Thebans. They would have had to lead them there, as they would not have found it alone. Anecdotal evidence indicates that the earliest settlement in the Ormonts Valley was at Vers l'Eglise. If it was a place that sheltered both Bagaudi and Thebans, it would explain why it is one of the coldest spots in the valley. It is hidden under the lee of the mountain, impossible to see even the smoke from their fires from any vantage point taken by soldiers hiking into the valley, and wild enough to deter an army.

The Bagaudi and others who lived high in the mountains overlooking the Upper Rhone Valley had kept at least part of their culture despite Roman law, systems of cultivation, cities and large mansions. Their lifestyle had evolved into that of seasonal pastoralists, in a large part of the high alpine country. It is not clear if they were Christians or that this was the basis for dissent against Roman rule, or whether they were simply protesting as people do who have been dispossessed of their lands and heritage, against tyranny by an occupying force.

Whatever the reason, this was a latent philosophy of a revolution against an imposed hegemony, emerging when conditions became too onerous. The Bagaudi did claim their ethnic and cultural differences, and their right to improve their social conditions. It appears that once again, climate variability also played a part. The dendrochronology data evidence indicates a cooling period beginning around 200 and lasting until around 315 AD. Rainfall in north-eastern France and central Europe became exceptionally variable from 250 until 650 AD, and a marked dry spell peaked around the year 300. The scientific data is in accord with written reports that there had been a worsening climate that had led to lower crop yields.

The name 'Bagaudi' appeared for the first time during the riots against Emperor Marcus Aurelius Carino around 282. Their first rebellion was quelled by the general Maximian, however the Bagaudi merely went underground, as dissent was recorded as returning several times up till around 460 AD. There is an account that during the decline and fall of the Roman Empire, the Bagaudi, in 407 forced general Saro to pay them all the booty collected in the countryside of Gaul in exchange for a safe passage through the Alpine passes. This indicates that the Bagaudi in Valais and the Chablais never lost the command of their own territory, simply migrated upwards to maintain their autonomy. It was a movement that spread throughout the formerly Celtic territories, and may indicate that the old allegiances were not dead. It is probable that there was ongoing communication between local people from Upper Rhone Valley at least with people in Gaul and possibly as distant as Iberia.

Roman control of most of Switzerland ceased in 401 AD. It may well be that in the wake of the departing Romans, the people sought to reestablish the old alliances of language and culture. However the tide of history was against them. After the crossing of the Rhine in 406, Switzerland began to be occupied by an organised invasion of the Burgundians, a Germanic people originating in Scandinavia, while the people, originally from the Helvetic and Rhone Valley tribes retreated further into the mountains. This period in history is often portrayed as a Germanic migration. To the locals it was just one more territorial incursion, an invasion imposed upon people who had not recovered their cultural identity after the Roman conquest. The Germanic tribes had after all aligned themselves with Rome to conquer the Gauls. And once again the next wave of incomers succeeded after aligning themselves again with Rome. The Burgundians engaged in many wars, and in 443 AD were given the territory of Savoie around Lyon by the Romans in the latter stages of the Roman Empire. The Burgundians of course proved to be treacherous to Rome in the dying stages of the empire, and through murderous plots and counterplots, the centre of power moved to Burgundy, where they managed for a time to take over most of modern Belgium, France, Germany and Switzerland.

Live by the sword, die by the sword, they were superseded by the Franks, a subsequent alliance of Germanic tribes.

Following the collapse of Rome in the West, the power vacuum was filled by the Frankish tribes united under the Merovingians. They succeeded in conquering most of Gaul in the 6th century. The Franks became very powerful. The Merovingian dynasty, descendants of the Salians, founded one of the Germanic monarchies that replaced the Western Roman Empire. The Frankish state consolidated its hold over large parts of Western Europe by the end of the eighth century, developing into the Carolingian Empire. Under the Carolingian kings, the feudal system proliferated. Monasteries and bishoprics became important bases for maintaining power and exploiting the local people. This empire would gradually evolve into the state of France and the Holy Roman Empire.

THE MORE THINGS CHANGE

The Eurasian Steppes before the last glaciation were dominated by pine forests. As the climate grew colder, spruce forests and tundra/steppe ecosystems developed, and anatomically modern humans hunted reindeer, horses and mammoths throughout the last glaciation, during the Upper Paleolithic as early as 40,000 BCE.

Language has played a dominant role in human activities since the last glaciation ended. Most of the European, Western and Central Asian, and Indian subcontinent languages have a common root stemming from a 'Proto Indo European' language. It is fascinating that somewhere in the last 10,000 years a common network of words, phrases and syntax developed, enough to make complex information understood. Analysis of PIE was accomplished by using techniques of linguistic reconstruction, including identifying common root words. This means of communication was possibly propagated in parallel with the adoption of agriculture 10,000 years ago, out of the civilizations of the Fertile Crescent, the Nile Valley and Nile Delta of northeast Africa, and the land in and around the Tigris and Euphrates rivers. One study using analysis of over 750 Neolithic sites found that the origins of agriculture were most likely to have occurred in the northern Levantine - Mesopotamia area. From this region, agriculture spread north westwards into Europe, also branching northwards around the western edge of the Black Sea. Agriculture was part of the pattern of technology innovation improving lifestyle. It led to population increases and pressure on available natural resources, and the migration and territorial expansion that facilitated the spread of agriculture to Europe.

One indigenous population, the Bug-Dniester culture, who lived around the Dniester River in the Ukraine, had an economy revolving around hunting and fishing. They were present in the region earlier than 6,000 BCE. A millennium later, they were at the frontier between the spread of farming and the aggressive expansion of the herdsmen of the Eurasian Steppes. It appears that farming was successfully adopted around the Dniester Estuary on the shores of the Black Sea. This would have reduced the opportunities for the indigenous population to conduct seasonal hunting. This period was a maximum warm period, when life was teeming, and the terrain biologically productive. As the farmers advanced, they either marginalised or converted local populations like the Bug-Dniester to an agricultural economy. The farming communities began to migrate towards the Eurasian steppes, once stable ecosystems. At the same time the pastoralism of the Western region of the Eurasian steppes was coming under environmental stresses, because of growing aridity. These people were also experiencing the patterns of population growth and restricted natural resources, including land availability. Their economy was centred on herding of domestic cattle, sheep and goats, a practice that started in the Fertile Crescent. Their traditional links to hunting and the horse, meant that they had become excellent equestrians. As ever, conflict began to break out as these populations sought to inhabit the same territory, and the original Bug-Dniester culture disappeared from view.

From a practical point of view, it is not surprising that the Ukraine Steppe was a frontier area. The Ukraine, even in the Bronze Age, was subject to territorial dispute. The lands around Odessa, the Dniester estuary and Lake Liman lie at the southern edge of the Ukraine Steppe. This period saw the advance of the Scythians, people who migrated back from the Altai Mountains, a journey of three thousand kilometres by way of Mongolia and Kazakhstan, with a culture developed around horses and equestrian skills. By the last millennium BCE, the Scythians had optimised their economy around the central Eurasian steppes. They were known to be early adopters and highly successful at mounted warfare, and most likely were not the first people to mount a hostile incursion into the territory around the Black Sea, but one in a long line of invaders. Part of the Indo European cultures that developed in the millennia after the last glaciation, the Scythians used the wheel and the horse, and their ability for transhumant herding (moving domestic herds from one place to another) to advance their culture and their territory.

By 3,000 BCE land in the region had become less suitable for agriculture, and the economy became less predictable, as hunting and herding once again became a dominant paradigm. Thus local populations became more nomadic and more decentralised, as the land became increasingly unsuitable for agriculture through decline in temperatures. Successive climate variations caused population migrations throughout the Bronze and Iron Ages until a wetter period around 1,500 years ago. Eventually the Scythians were conquered by both the Macedonians, and people from the Iranian Plateau.

U-Kraina (in some Slavic combination) means 'at the edge' or 'at the border' and was indeed named that way because much of its territory marked the boundary between the open steppe grasslands, and the forest steppe with its predominantly farmer settlers. Eventually this region was transformed by colonists from the south-west, (Ottoman, Tatar, Greeks and Romans) and from the north by Slavic peoples and the Verangians, Vikings who migrated south-west in search of trading opportunities.

The more things change, however, the more things stay the same, and evidently that ancient hunting and fishing practices did survive around Chabag and the Dniester, until Swiss colonists arrived in the early 19th century. To establish the fisheries on Lake Liman and the Dniester estuary, the local Turkish population had dug a number of canals extending inland 100 or so metres, and ending in a cul-de-sac. At spawning time many thousands of kiffali, a kind of Targe herring, would enter the canals to spawn and were easily netted. Some were eaten fresh but most were salted down in barrels and stored in cellars. This provided the settlers with a most acceptable and cheap variation of diet. The sandy soils near the Black Sea produced big crops of watermelons and rockmelons in the 19th century.

The big game of the locality consisted of wolves in winter plus a few bears, whilst on the steppes there were foxes and hares as well as partridge, quail and bustard. In the swamps, in addition to myriads of mosquitoes, there were wild swans, geese, ducks of many species, plovers and other wildfowl. Shooting was the chief sport of the Swiss colonists at Chabag.

The Roman Empire reached its greatest territorial extent during the reign of Trajan (53- 117 AD). They conquered Dacia from 100-107 AD. Where the imperial boundary of that period traversed the Russian plain, they caused a big moat to be dug, strengthened militarily every few hundred yards by crenellated towers or fortlets, thus somewhat resembling Hadrian's Wall at Carlisle. Chabag was right on this old boundary line as it meandered westerly from the Black Sea to the Carpathians.

The towers had eventually disappeared centuries before their arrival, but the Swiss settlers levelled off the moat in 1823 in order to lay down their streets and build their homes. Beyond the village, even in 1887, the colonists could easily follow the line of Trajan's boundary, as plainly definable as the Great Wall of China. They restored some of the derelict Turkish vineyards and planted many new ones on more modern lines. Windbreaks were a necessity against the strong steppes winds and those from the sea. The vineyards were bordered with black and white mulberries, walnuts, acacias, apricots and other suitable trees.

The current territorial incursions into the Ukraine and the aggression from the north by the Russians seems like a sad echo from an ancient past.

ANCESTRAL JOURNEYS

Can we abandon senseless hostilities and the stupidity of wars which benefit no-one, especially when we know that the climate systems are on the edge of irreversible unpredictability? What does it take for humanity to lay down weapons and pick up reason and reconnect to nature, the wellspring of our existence? What does it take to recognize that war is an ignorant denial of human culture, particularly at a time when population growth, natural resource exhaustion and climate change are risks and threats to all people of every nation?

What lessons can be learned from the ancestors' long, slow journey from Africa to the Fertile Crescent, and the drift westward to Europe, overseas to Britain and Ireland, and to Switzerland over the alpine passes? Clearly climate change, population growth and limited natural resources were at the root of our amazing extended, adventurous voyage, our collective odyssey. The pattern has repeated itself at least throughout the last sixty thousand years. As we have spread throughout the planet, the need to migrate onwards has become increasingly pressing. By the eighteenth and nineteenth centuries, when available ancestral land in Europe offered limited opportunities, in the face of population growth and dwindling resources, there was another large-scale migration in search of new opportunity to the new world continents. At this time, improved transport meant that the whole process could happen on a bigger, grander scale, and so much faster than the slow movements and incursions of the past. We have to recognize that population migration displaces existing populations, and destabilises local ecosystems during the transition phase.

Firstly, who cannot but feel awe for the temerity and perseverance of the ancestors in the face of adversity? Next, we clearly have been very intelligent people for a very long time, making the best use of resources that were available to us, and readily turning our brains to inventing new technologies to meet new challenges. This clearly contradicts the peculiar slant populist literature ascribes to history and the ways of the past. For example, the people of ancient Gaul, clearly purveyors of a superior social order, with advanced knowledge of astronomy and navigation, art and music, education and oral literary traditions, have been portrayed as barbarians for two thousand years, on the dictates of their conquerors, the Romans. The alignments of their buildings were for practical purposes of terrestrial route finding on a grand scale, their local maps a record of land use and a working knowledge of best use of local resources. And yet we have mythologized the Romans with the slavish devotion of the conquered, repeating the ill-informed Roman historians for over two thousand years.

For the past two thousand years, we have been dismissive of our connection to the plants, animals, geological formations, and the accompanying ability to live sustainable lifestyles, while lauding military might, conquest and monetary wealth. We owe our current benefits, blessings and standards of living to the ancestors, who had the courage to ignore the dominant paradigms and populist values. It was the technologists, the scientists, the thinkers, the artists, the writers, innovators and the heretics who blazed the trails for our new directions, not the princelings, merchants, warriors and politicians. It is our ability to innovate, our energy and industriousness in the development of new technologies, balanced by a proper appreciation of music and the arts that are our strengths. The planet is politically and economically at a point of international crisis. To protect our civilization, we have to take carefully calculated risks for managing people and economies, with a proper value placed on natural resources. We have to maintain a culture of being able to speak our truths in public without fear of reprisals to help us to solve our current problems, which are unprecedented in scale, even as they repeat the problems of the past.

We have to communicate freely, in reasoned discussion amongst ourselves to develop new ways of addressing problems that clearly have not been solved by increased regulation and decreasing civil liberties. We have to put forward our ideas and test them amongst our peers to find our best next moves. We have to be open to new ideas, and willing to make changes in our own lives, and to support carefully planned transformation programs collectively, not only at a national level, but globally. We have to ensure that the United Nations reflects a pluralistic, inclusive approach to the urgent problems of sea-level rises, and changing weather patterns, equally working for both wealthy and fragile nation states, as climate change does not discriminate between wealthy and fragile countries.

We have to make the social and technology changes, not from a standpoint of nationalistic gain and empire-building, rather from a shared perspective of working to improve economies in harmony with the planet's biosphere. Particularly, we have to embrace the major challenges of responding to climate change with all the technology and expertise available to us. This means transitioning from a status quo that is based on old economic values and unfair exploitation of people and local environments, to injecting capital and investing effort and energy into innovation with clean, sustainable technologies in all industries. Another important development is the placing of value on the planet's resources of water, air and sustainable terrain. We also have to put a fair value on labour, paid and unpaid. The current excessive salaries of those who can gouge short term profits from the world's common resources must stop. We have to extend our initial foray into automation using computing power with creative and imaginative solutions in all fields of technological advancement.

Formal engagement in transformation of the financial system is a necessity. Short term profits for shareholders at any cost serves nobody. What is required is a system where not only is a value placed on all of the planet's natural resources, but also a negative value placed on greenhouse gas emissions. We have to establish positive global markets to encourage sustainability of natural resources. We have to accelerate the valuation of women's contributions to economic growth, particularly for those in rural communities. We have to protect and preserve all existing animal and plant species, enabling the wild species to continue to live in their native habitats, supported by the natural environment. We have to do our best to conserve all the natural resources and assets that will support sustainable growth in every local community. And finally we have to address the discrepancies of the commonwealth of clean air, fresh water, and healthy land management, so that decision making that affects us all can no longer be concentrated in the hands of a few individuals by virtue of the amount of money they have managed to amass, usually at someone else's expense.

The 19th and 20th centuries were characterised by people making long journeys in search of opportunity, migrating en masse to find a better standard of living in another part of the world. And yet the same themes have been at play since the end of the last glaciation. These themes are displacement of people by the diminution of natural resources in their native regions, and the use of force, war and conquest to secure those natural resources by incomers, coupled with increases in population. The trend is accelerating in the 21st century, as the pressure of population on natural resources, particularly in fragile countries, continues to increase. Today there are no untapped locations left in the world to colonise and populate, as the world's natural resources of silviculture, agriculture and aquaculture, as well as water, food and energy supplies for rural areas, villages, towns and cities are already stretched to the limit. Clearly something has to change.

The problem is not the availability of natural resources. There are also plenty of people with the knowledge, skills, experience and wisdom to manage those resources. Renewable energy, sustainable food production, clean technology, land management, and low carbon building systems are all mature, or well into development cycles for innovative practices. Knowledge about forests, oceans and wilderness is freely available, with many organisations offering excellent services for conservation and rehabilitation. Science and technology research into positive, resource friendly industries continue to be wildly successful. Information technology cloud networks are readily available to share public and private knowledge across national boundaries and borders, across scientific and economic disciplines.

So what is the problem? We are well on the road to using our new found knowledge and technology to sustainably manage the planet in sympathetic harmony with the rest of the living beings, both fauna and flora that have evolved alongside us. Can we adopt the role of custodians of the beautiful and rich forms of life that have nurtured our evolutionary development as human beings? Really, can we evolve from our past bad habits of force and enslavement to acquire resources, treating the natural environment as an inexhaustible supply, in time to stabilise Earth's biosphere before our weather systems and climate change more dramatically than our current civilizations can manage?

We have to recognize that now is the last opportunity we have to learn from past mistakes, particularly insufficient planning to prevent exhaustion of natural resources, and the management of greenhouse gas emissions directly causing global temperature rises. Engaging in warfare and conquest has never solved the problems of declining resources. Becoming superstitious and fearful as a result of facing the horrendous consequences of losing a war does nothing for progress. Becoming cruel and power obsessed as a result of winning a war does nothing for humanity. These responses, learned from our parents, and their parents, take us back in time to the original events that produced the fear and flight behaviours. We have to change these patterns, break the cycle, and substitute reason, compassion, wisdom and knowledge, to take positive action for our own and Earth's wellbeing.

Time to put political pressure on our governments to comply with the United Nations Framework Convention on Climate Change. The UNFCCC requires every country to lodge their plans for responding to climate change. We have a job to do ensuring that the greenhouse emissions estimates are accurate, that countries develop and adhere to real strategies for reducing emissions, using the best possible technologies. We have to ensure that individuals and organisations provide the necessary investment to encourage substantial growth in clean technology industries and jobs at every level in every community. We have to demand that our governments provide the support and the legal framework to mobilise the money to address the problems of human induced climate change. And we have to ensure that these activities are in a governance framework that discourages the inevitable attempts by some at corruption and greenwashing.

PART 2 - RESPONSE TO CLIMATE CHANGE

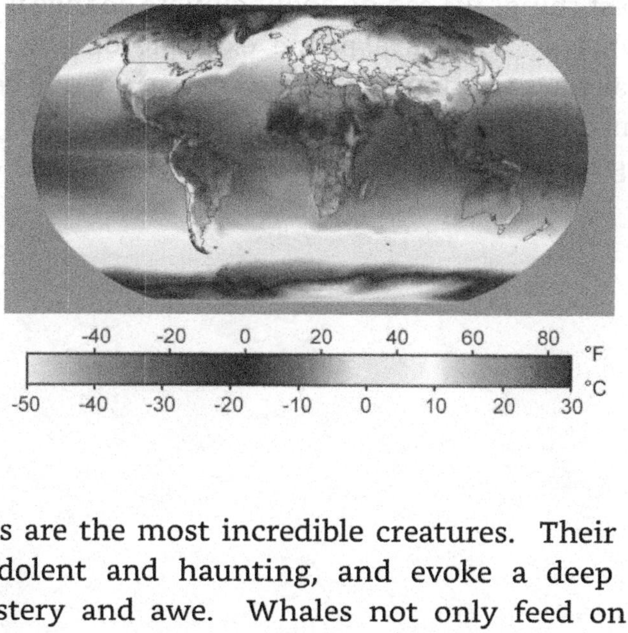

Whales are the most incredible creatures. Their songs are redolent and haunting, and evoke a deep sense of mystery and awe. Whales not only feed on krill, they keep the whole photosynthesis chain alive in an ecological phenomenon known as trophic cascade. Trophic cascades occur when a species in a food web interact with the behaviours of the species on which they feed. Whales sustain the entire living system of the ocean. They eat the krill in the depths, then return to the surface where they fertilise the ocean with massive quantities of iron and nitrogen into the photic zone, providing nourishment for more plant life. They engage in a constant recycling of nutrients from the depths to the surface. Whales accomplish more than all the wind, waves and tides put together to encourage the growth of plants in the seas, which means plenty of fish, as they feed on the plants. An abundance of plants means that much more CO2 is absorbed from the atmosphere, decreasing the global warming caused by greenhouse emissions. The whale population is seriously endangered, so one of the simplest things we can do about climate change is saving the whales. This is a prime example of how simple natural ways to address climate change are our best bet. Preventing further acidification of the oceans is key to managing the risks of runaway warming effects.

CLIMATE AND WEATHER

Climate is usually described in terms of the mean variability of temperature, rainfall, snow, and wind over time. Earth's climate system evolves because of internal dynamics of atmosphere, ocean currents, weather, as well as external factors at play, including volcanoes, changes in the planet's orbit, and solar activity from sunspots. It is only in the past century that human activity has had a marked effect on climate. Anthropogenic (human induced) climate change is the only one about which we can have any influence.

The wheels of civilization have turned on polite conversation about the weather. What is weather, how does it work, and what is its relationship to climate? For us humans, weather is the daily experience of sun, rain, wind, snow, humidity etc. Climate is a longer term overview of the cumulative effects of our daily weather events. Climate zones have similar characteristics, for example hot and dry, warm and wet. As well as describing the effects of weather, our experience of climate has developed from interactions between the planet's systems, such as geographic features, for example mountains and plains, ocean currents, ice sheets, atmospheric air movements and the biosphere, the living organisms.

Water is the key element and generator for life, and it circulates constantly on Earth's surface, and in the atmosphere, influenced by many factors. One of those is human activities. Water feedback loops are key to the Earth's biosphere maintaining a stable state suitable for life. We can all experience and recognize simple feedback loops in the weather. If there is enough water vapour in the clouds, it rains. Of course climate systems are not so simple. There are three key feedbacks on Planet Earth, interacting with one another to determine our weather. They are water, ice and solar radiation. Our simple example is solar radiation heating the water vapour in the clouds, which in turn reflects more heat back to the surface, which increases the evaporation. All complex systems are made up of a number of simple parts.

The climate is also constantly affected by heat radiation, as bodies such as the land and the ocean, having warmed up, try to cool down again. The level of radiation reaching us determines average global temperatures. While greenhouse gases provide us with a stable range of weather conditions and climate that support life, you can have too much of a good thing. If temperatures on the planet get too hot or too cold, human beings cannot thrive. Currently the level of greenhouse emissions from our activities is allowing the atmosphere to radiate more heat towards the surface of the planet. And this is having a very noticeable effect on the cryosphere, Earth's frozen regions.

Our Celtic ancestors were right to focus so much attention on the sun. The Earth's climate has changed throughout history. The ending of the last ice age about 12,000 years ago marked the beginning of the modern climate era. Most climate changes are attributed to very small variations in Earth's orbit that change the amount of solar radiation energy that is experienced by the planet's biosphere. There have been at least five major ice ages. Each glacial period was subject to positive feedback increasing the severity, and negative feedback, eventually ending it. The weight of the ice sheets was so great that they deformed the Earth's crust and mantle. As the ice sheets melted, the ice-covered land rebounded, albeit at a very slow rate, causing instability and tremors, a process that is still in action today. The great melting events also changed the gravity and wobble of the Earth's rotation.

So what do we know about the most recent of Earth's great climate variations? Global warming started 15,000 years ago. By 5,000 BC, the last of the Northern European ice sheets had disappeared from Scandinavia, and the last of the North American ice sheets were gone from eastern Canada. As yet the causes are not fully determined, major factors are suspected to be changes in the Earth's orbit, shifts in the tectonic plates, fluctuations in ocean currents, and last but by no means least, the volume of greenhouse gases in the atmosphere. One theory has it that glaciation starts and ends because of increases and decreases in CO2 levels in the atmosphere. There is no doubt that the levels of atmospheric greenhouse gases are dependent variables in relation to Earth's temperature variation. Today it is unequivocally human activity that is increasing these levels. More than 800 authors, selected from around 3,000 nominations, were involved in writing the latest IPCC AR5 climate change report. Rapidly increasing industrial activities accompanied by huge growth in the number of people on the planet has been the trigger for the large acceleration of gas concentration seen in recent decades, and we are now in territory where we have never been before.

The United Nations Intergovernmental Panel on Climate Change (IPCC) defines radiative forcing (RF) as a measure of the influence particular events have in altering the balance of incoming and outgoing energy on Earth. RF measures the amount of heat trapped, either in the atmosphere or on Earth's surface, and provides a way to view the factors that influence climate change. While climate variability is normal, from a historical trend perspective, since the end of the 19th century average global land and ocean temperatures have risen at a rapid rate. One of the biggest causes of sea-level rises is the net melting of the cryosphere, resulting in loss of mass of the world's glaciers and the Greenland ice sheet at an unprecedented rate. The biggest anthropogenic (caused by human activity) source of greenhouse gas emissions is the burning of fossil fuels, oil, coal and gas.

Projections of climate change through the rest of the century show amplified warming in the Arctic compared to the rest of the planet. Sea ice loss plays a strong role. With less ice in spring and summer, the water near the surface of the ocean gains more heat through absorption of solar radiation. The cause and effect can be simply described: the smaller the surface area of reflective white ice, either because it has melted, or because it is discoloured from pollution, the more heat is trapped by the surface of the Earth, the faster the ice melts. This is another positive feedback loop affected by other climate interaction variables, such as ocean and wind currents.

Climate feedback loops are behavioural phenomena. At some point in time, previously predictable processes can become uncertain, resulting in runaway changes, and accompanied by unknown effects on climate systems, except that they can result in swings in range of temperatures, fire, and changes in rainfall, snowfall, and tempest events, causing wild weather patterns.

One natural factor affecting weather events that is not entirely understood is known as the Arctic Oscillation, an atmospheric circulation pattern in which the atmospheric pressure over the polar regions varies counter to that of the middle latitudes. The differential between high Arctic and low middle latitude atmospheric pressure results in warmer weather over the northern polar region, and cold stormy weather over the temperate areas where people live, with dramatic weather events such as experienced in 2009-2010 in Europe, the United States, and Asia.

The IPCC states that to keep the level of emissions to current levels, countries have to make 'significant cuts' by 2020, and 'substantial cuts' to greenhouse emissions by the middle of the century. When temperatures rise, ice melts, sea levels rise, and weather instabilities can cause erratic instances of extreme rainfall, winds, tides and storm surges. Without dramatic changes to the world's reliance on fossil fuels, Planet Earth is set to become a much less hospitable place for human beings, even hostile to our existence. The current emissions levels mean that it is already too late to stop the rise in sea levels because of melting ice. This phenomenon is now underway, and because of the lag between any reduction in temperature, and the slowing of the melting of ice sheets and glaciers, the best case scenario is that within a couple of centuries, the sea level rises stop. We are playing with forces that we do not understand, and all the warning signs are there that, like the sorcerer's apprentice, our activities meddling in climate variability are having unfortunate and unpredictable consequences.

Our collective use of energy is at the heart of climate stability. Yet, as a species, we are still struggling to find the collective political will to forsake fossil fuels, the biggest source of greenhouse emissions, which were formed during a great mass extinction event. Why it is so difficult to interest the finance industry in a rapid response to capitalise the development of technologies which are cleaner, cleverer, greener and now cheaper than coal, oil and gas based energy production currently, is a mystery. A rational and reasoned response would have investors queuing to scrutinise opportunities for clean technology, low carbon supply chains, and a carbon market with a price that reflects the real value of curbing emissions to the global economic outlook.

While United Nations programs start to challenge our collective lack of action, and countries pay lip service to mitigation activities, the level of political interference and the potential for greenwashing and corruption is disturbing. We all, individuals and corporations have to take responsibility, demand our politicians support a low carbon economy, and seriously engage in energy efficiency. We also must provide leadership for others to do likewise. As a species, we have no choice but to curb our addiction to money as an end in itself, rather than a means to an end, if we are to take effective action in time to give our descendants a decent future. The burning question is can we change the influence of the finance industry on our politicians, to abandon the paradigm of shareholder profits at any cost in time? The ancestors must not only be turning in their graves with uneasiness over our current commitment to our collective future, they must be getting ready to rattle their bones.

Civilizations have risen and fallen in response to benign and difficult climatic conditions in localised regions. The ancient civilizations of Persia, Mesopotamia, and the Aegean civilizations such as Greece all experienced a decline and fall that was a result of a domino effect from climate change causing territorial aggression. Agricultural settlements made prime targets, while city states failed to recognize that both over-population and injudicious use of natural resources would be their downfall. It is around 5,000 years since this phenomenon commenced. Are we slow learners?

The current situation in which we humans find ourselves is an incredible challenge to re-establish harmony with the natural environment, to sustain our existence. Although in the past we have always responded with a new round of technology innovation, never before have we been in danger of using up the planet's entire stock of forests, the soils, the waterways, oceans, and even the air that we breathe. Our collective recognition of our predicament, and concerted action is urgently required to preserve ourselves, the animals, plants, oceans, skies, mountains, rivers and terrain in a way that is life enhancing. In fact, we all have to take community action to effect global change for the good, and unless we can stop our unfortunate habit of going to war to solve problems, we can expect a spiralling loss of control of balance between humans and nature. Otherwise we cannot expect to thrive, or perhaps even survive as a species. We only have ourselves to blame, and we are the only ones who can affect the necessary behavioural changes in our communities, corporations, governments and social groups to make sustainable development the dominant paradigm, and to curb excessive and dangerous emissions of greenhouse gases.

Chaos theory and evolutionary biology indicate that our blue planet exists in a form of gravitational and climatic equilibrium that can experience monumental changes from a single initial event, (though not quite the flap of the butterfly's wing). These changes occur because, in the scheme of things, the range of temperature to support the living systems of the biosphere is a relatively fragile balance. This balance is governed by the same fractal patterns that form and conform to the laws of the spatial and temporal material universe in which we are inhabitants. We may be getting extremely accurate at recording weather, temperature and precipitation, metrics which show us that temperatures and sea levels are rising. However presently, we do not have a supercomputer powerful enough to predict in a timely way, the outcomes of playing with an already fragile environment when the patterns of acidification of the oceans, polar ice, fires and floods caused by extreme weather, are becoming increasingly unstable.

Local weather around the globe is becoming more difficult to forecast. It has happened in the past. A single change, such as the drying up of a water source, or the disappearance of a species on which a number of others depend could be enough to resonate locally, to initiate increasingly dramatic swings that make life too difficult for some species. We are already experiencing an alarming increase in the rate of extinction of flora and fauna. This comprises a net unknown effect on Earth's stable biosphere. We simply do not know what, where, how and why extreme weather is going to happen. Like the oscillations of a weight suspended from a rope dangling from a helicopter – as the arc of the swing gets wider, any sudden motion can cause wild movements in unpredictable directions when systems reach a 'tipping point' of instability. We currently do not have the tools or knowledge to predict how, when, and where the breakdowns occur.

We live in a complex set of conditions that amazingly form a wonderful, stable nourishing environment to support water based life. History demonstrates that we can lose these conditions all too readily. When the planet suffers one too many destabilising events, the speed of change is rapid and uncompromising. The Earth seeks to rebalance the harmony of its biosphere, and in so doing can dispense with any elements that compromise the limit of its systems which provide us human beings with safety and security. Truly our blue planet is the Gaia, a single living organism of which we humans, and the rest of life on the planet are the biological cells. Life evolved in such a way that single celled bacteria learned to co-operate and form biofilm communities to achieve sophisticated functions by specialist cells. By nature, we and the Earth are an integral part of this evolutionary universe, governed by spin-centric energy that has created matter through aeons of ordered and chaotic interaction with the original energy plasma, an amazing phenomenon that eventually gave us our solar system and Planet Earth.

The evolution of our universe is a delicate balance of expansion and gravity, in which domains of less entropic, quantum improbable ordered events time our existence, from an initial condition. Change is the normal state of affairs. A change to a fundamental energy state over time gestated the elements of the periodic table, forming matter, then eventually, life as we know it over billions of years. And this process continues today. This is the cosmos, where we inhabit a tiny piece of rock, in a galaxy of billions of stars, part of a galaxy supercluster binding together with regional galaxy superclusters, all rotating and travelling through space and time, all governed by the laws of the universe, which we do not in any way, shape or form fully comprehend. We are part of it, a part that has acquired a certain facility with technology. Our successes are manifold, however our propensity for turning our gifts for innovation towards warfare and biosphere destruction proves the axiom 'a little knowledge is a dangerous thing'.

The basic science of global warming is the way that radiation from the sun is either reflected back into space from the planet's surface, or remains 'trapped' in the atmosphere long enough to have an effect on the planet's climate. There are limits to stable systems, including Earth's orbit, because of some fundamental laws of nature. Earthquakes obey these laws, and so does the stock exchange. It must then come as no surprise that weather and climate systems are subject to unpredictable fluctuations. The mathematical theory that attempts to describe the laws of these changes is called 'Chaos Theory'. Simply put, this theory tells us that for a small change in the initial state of a system, when there are feedback loops present, the system in question can become unstable in unpredictable directions at times that cannot be precisely predicted.

CLIMATE STABILITY

The 'greenhouse effect' is the fact that increasing the level of particular gases, such as carbon dioxide and methane in Earth's atmosphere, causes more radiation to be retained and scattered, including towards the planet's surface. It is not hard to understand intuitively why ice and snow are the most effective reflectors. One of the most worrying feedback loops in relation to greenhouse gases and climate change, is that the more the ice sheets and glaciers melt, the less ability the Earth has to reflect heat from the sun back into space. As well as the obvious consequences of global warming, the effects of melting the planet's ice covers are not fully understood.

Despite the fact that unpredictability happens at the edge of chaos, our solar system has proved to be very reliable in rebalancing itself over time. Sometimes it is thrown off-balance by events, however it eventually re-establishes the balance of the biosphere to support liquid water and carbon based life. The more we understand about the key principles of climate change, and how the weather is generated from climate systems, feedbacks, cycles and self-regulation, the more it seems that in our own best interests, we do not want to mess around with the balance of the natural cycles of living systems here on Earth.

Studies of climate change in the very ancient past, as well as the recent past, indicate that carbon cycle change may be a key indicator of transition and significant change to the climate. We may be hearing an alarm to which it is not yet too late to respond, even though it is getting very close to midnight.

Meteorological offices around the world collect data from physical atmospheric and oceanic variables to understand and monitor global climate variability and change. While signs of climate change are the accurate measurements that have identified trends of rising land and ocean temperatures and rising sea levels, this is accompanied by changes to the global carbon cycle. Since the Industrial Revolution we have been burning fossil fuels on a grand scale. More than 80% of the increases in the human component of global warming are due to combustion of oil, coal and natural gas, the fossil fuels. The other major problem is the deforestation of the planet, particularly the rainforests in the tropical north of South America, and SouthEast Asia. The link has been firmly established between changes to the climate and the global carbon cycle. Not all of the carbon ends up in the atmosphere, large quantities are being deposited on land, and a large part of the carbon ends up being deposited in the deep ocean, which has become a carbon sink. What is not known is what happens to the biosphere as we pump more and more carbon into the planet's climate systems.

The IPCC (Intergovernmental Panel on Climate Change) is a body comprising thousands of scientists and other experts who contribute to writing and reviewing reports. In the latest, the Fifth Assessment, the IPCC has developed and published four greenhouse gas concentration trajectories. They describe four possible climate futures, all of which are considered possible depending on the quantity of greenhouse gases emitted in the years to come. They model a range resulting in possible global mean temperature rises between around 1.5 °C and 5 °C, from 100% replacement of fossil fuels to business as usual. The report also documents the likely changes to weather, air quality, ice and snow cover, and ocean acidification across the range of the scenarios.

The projected impacts for business as usual are catastrophic in terms of extreme weather events. Although from the least to the greatest change to the carbon cycle, current forecasts project irreversible warming over the next couple of centuries. Limiting the warming caused by anthropogenic emissions is going to reduce the impacts on land and ocean acidification. In any event it is virtually certain that global mean sea level rise will continue beyond 2100, with sea level rise due to thermal expansion set to continue for many centuries. At stake is our ability to thrive and even survive with undisputed scientific analysis predicting a future of between 2 °C and 4 °C global mean temperature increases above pre-industrial levels, for the path we are most likely to take. These rises are accompanied by associated sea level rises with a probable range between 0.26 and 0.82 metres, although some scenario estimates are being set higher at 1.9 metres.

Change to the carbon cycle is going to have a large effect on human health and the built environment, as changing weather patterns are having devastating effects on some communities. Inundation of houses, erosion of coastal zones and increasing wildfires are extreme situations requiring risk and hazard reduction. For some island nations of the Pacific, entire communities are threatened by rising seas. Many coastal regions in every region of the planet where events such as hurricanes, cyclones and monsoons occur, are susceptible to loss and damage from storm surges, which have been responsible for many deaths over the past decade.

Climate change is going to play a big role in food security for the more than seven billion people living on the planet. Heat waves, drought, and flooding constitute potential global food security issues, as we have to adapt to different weather patterns in the food producing regions. Every part of the globe is vulnerable in both wealthy and fragile countries, to changing conditions for food availability and accessibility, as well as food distribution stability. We have to persuade our governments to support and fund innovation in agribusiness to reduce toxicity from food production, and increase carbon sequestration with our choice of food crops.

We know there are certain changes, disruptions, and risks of disasters occurring that have to be addressed. There is a strong probability that we are already experiencing increasingly chaotic weather patterns in many parts of the globe as a direct result of feedback loops triggered by changes to the carbon cycle.

James Lovelock's Gaia theory examines the evidence for the equilibrium of living systems that over the history of the planet, engage in a perpetual balancing of this self-stabilising energy biosphere. Proof of microorganisms living on Earth during at least the last 3.2 billion years gives the theory a very long period for demonstration of this observed homeostasis of the atmospheric biosphere, as it self-regulates to maintain life. This is an absolutely amazing observation, as during this period, the level of radiation from the Sun has increased dramatically. The sun has got hotter. In response to the associated changes in temperature, and the level of the carbon, hydrogen, oxygen and nitrogen (CHON) based gases in Earth's atmosphere, somehow the conditions have remained within the bounds that support organic life. The inference is that there is some mechanism that controls the balance of atmospheric gases of the planet, with the express purpose of maintaining itself. And the exciting conclusion that one inescapably draws from this theory is that we are living in an innately intelligent universe, manifest in the energy patterns underlying the physical structure of matter of which Planet Earth, our blue planet, our only home, is a living example.

While Lovelock and his colleagues have explored the chemical composition of the atmosphere in relation to the role of thermodynamics and decreasing entropy (disorder) in the energy of living systems, they acknowledge that complex systems, made of a community of simple parts, like the atmosphere, or a rainforest, appear to behave as intelligent cells in the whole biosphere of life on Earth. Feedback loops are the mechanism by which the planet regulates the temperature range. The energy from the sun, causing heat on Earth, has a certain range of stability, maintained by elements such as the clouds, and the gases in the atmosphere. The net effect of all the positive and negative feedbacks that have an influence on how energy is radiated, governs the ability of the Earth to mirror energy back into space. The albedo index measures how well ice, clouds, water vapour, the ocean and different types of terrain reflect radiation, which is otherwise converted into heat. The heat range of course has a direct effect on which living systems are able to survive the climatic conditions. While simpler organisms, like bacteria can function in a wider range of temperatures, we humans, having a complex biology, although made of simple parts, have quite a narrow range in which our bodies can function.

While the science of climate change is fascinating, current thinking is that natural systems hold the key to our best chance of maintaining the stability of climate systems to suit life on Earth as we know it. More than 10% of greenhouse gas emissions are coming from the clearing and burning of forests. Greenhouse gases stored in peatlands and the tundra are beginning to be released by thawing of permafrost, while soils are being depleted, degrading their ability to store carbon. Natural carbon capture and storage holds the most promise for stabilising and limiting the level of greenhouse gases blanketing the planet. (And as in most things, the wheel turns full circle. The sixties and seventies saw the original mass movement of concern for the natural environment. Two of the most significant campaigns from this era were to save the whales and save the old-growth forests. Interestingly enough, both of these endeavours are key parts of the chain of natural processing of greenhouse gases on a large scale.)

It is well-known that forests achieve a balance between the amount of carbon dioxide absorbed by growing trees and plants and the amount of CO2 released back into the atmosphere by the decomposition of forest debris. Old growth forests are considered to be those that have remained untouched for at least a hundred years. The relatively small remaining stands of old-growth forests are a global terrestrial carbon sink. In fact, not only do old trees continue to store carbon in their wood, forest soils also appear to be actively capturing carbon over time. Old-growth forests are biologically diverse, and home to many rare and endangered species of plants and animals, which also play a vital part in maintaining the stability of the biosphere. And they are incredibly beautiful. The sensation of standing within a rainforest is one of healing, tranquillity and energy radiance, once experienced, never forgotten. Old growth forests accumulate carbon for centuries and contain large quantities of it, however much of this carbon, even soil carbon, will move back to the atmosphere if these forests are disturbed. One of the simplest ways to reduce greenhouse emissions is to prevent logging in the Amazon, Indonesia and Malaysia, and Tasmania. These trees are the lungs of the world. The atmosphere of the Earth has a different composition from that of other planets in part due to the biochemical reactions with the plants. Forests pump oxygen into the air so we can breathe. Saving the forests is a given. (Recycled paper is all too often from plantation timber that is grown where old growth forests were destroyed, forests are a prime target for greenwashing).

There are large risks to food security globally and even more devastating, reduction in surface water and groundwater resources in most dry subtropical regions. It is very likely that heat waves will occur more often and last longer, and that extreme precipitation events will become more intense and frequent in many regions. One of the most vulnerable terrains to rising sea levels is the tropical coastline, where deforestation and removal of mangroves, a vital fish breeding ground, has occurred. Projects for restoration of mangroves and control of erosion and sea level rises, for example in Indonesia and Malaysia, have an excellent effect on local ecosystems. Economic activities for local people, generated by mangrove protection, include environmental management and generation of sustainable commercial agribusiness.

While the existing conditions are understood in terms of stable past climates, more knowledge is required to manage the changes in the local ecosystems that have been caused by variations in temperature ranges, weather, tidal and wave patterns, affecting the topsoil, coastal erosion and sedimentation.

GREENHOUSE GAS EMISSIONS

Greenhouse gases work a little like a greenhouse, however the glass in this case is a layer of gases. The atmosphere traps gases, which form a blanket around our planet. It is the composition of gases that determines whether radiation reaches the planet's surfaces or is reflected back to space. The levels of water vapour, methane, carbon dioxide, nitrous oxide and other gases determine whether the average temperature of the Earth's surface rises or falls.

Global-warming potential (GWP) is a relative measure of how much heat is released into the atmosphere by the greenhouse gases, of which methane is the second most prevalent. Although the level of methane is only around 1.8 ppm, in terms of global warming potential (GWP) it is around 29 times as potent, per unit of mass, as carbon dioxide (CO2) in raising atmospheric temperatures. A vast expanse of permafrost in Siberia and Alaska, a permanently frozen mass as the name suggests, a storehouse of methane and ice, has started to thaw for the first time since it formed 11,000 years ago. The term

CO2e, or carbon dioxide equivalent, refers to the whole suite of greenhouse gas emissions standardised in terms of Global Warming Potential (GWP).

What gets measured gets managed. Under the Kyoto Protocol, countries' actual emissions have to be monitored and precise records kept of any carbon trading. The UN Climate Change Secretariat, based in Bonn, Germany, maintains a registry to verify that transactions are consistent with the rules of the Protocol, and countries have to submit annual emission inventories and national reports at regular intervals. The current agreement has flaws in its structure. It excluded the conservation of old growth forests and did not make provision for land use changes, which would place a value on increasing carbon sequestration. Increases in fossil fuel burning by major polluters could be traded advantageously for projects that are by no means equivalent in Global Warming Potential (GWP). And lastly, countries are estimating their carbon accounts with varying degrees of accuracy, meaning that a carbon cap and trade market is not yet a viable mechanism for lowering emissions. Everybody knows that markets do not like uncertainty. Wildly different levels of accuracy across industry sectors and countries are never going to provide the market mechanism that would result in stable and successful reductions in greenhouse gases.

A new agreement is being formed, forged and negotiated by the UN FCCC, to be signed in Paris in December 2015. The agreement is being negotiated by the major polluting nations such as China, US, India and Russia, equally with the small island states in Micronesia, the Pacific, the Caribbean and the Indian Ocean that face serious threat of permanent inundation from sea level rises. Naturally there has been a veritable frenzy of tradeoffs, backroom deals, political posturing and cartels of vested interests over the past couple of years aiming to get an advantage from the new protocol.

By stimulating public interest, online monitoring, reporting and comparative analysis can play a vital role in providing a governance mechanism for measuring the real reduction in national inventories of greenhouse gas emissions. The largest source of emissions is currently the burning of coal, natural gas, and oil for electricity, both for domestic and industrial consumption. With political will, electricity can be metered and managed on an industrial scale for national electricity grids. There is no technology bar to doing so.

Land use, including deforestation, land clearing for agriculture, fires, burning peat, and releasing methane from permafrost are serious environmental factors that cause direct emissions. Agricultural emissions result largely from the management of depleted soils, livestock, rice production, and biomass burning. Over 90% of the world's transportation energy comes from petroleum based fuels, largely gasoline and diesel. Fossil fuels are burned for all forms of road, rail, air, and marine transportation. The construction industry contribution to global warming, other than indirect emissions from electricity and transport, is mainly from the manufacture of Portland type cement and cement products, because of the very energy intensive manufacturing processes. In the waste sector, landfill and wastewater are responsible for the largest source of methane (CH_4), and nitrous oxide (N_2O).

Telemetry systems, public and private information technology cloud systems and compulsory online published reporting by all countries can readily provide a level of data accuracy that is currently missing. Information and Communications Technology (ICT) systems can provide near real-time analytics and operational intelligence of emissions reduction from electricity usage, enabling an accurate set of metrics on which to base a cap and trade carbon market. This would give investors and traders the confidence they currently lack to support global trading in carbon futures.

LOW CARBON AGRIBUSINESS

The Intergovernmental Panel on Climate Change made the following key findings on the effects of climate change on farmers worldwide.

'Climate-related effects are already being felt in many regions, with harvests of staple crops reflecting reduced productivity of soils and changing rainfalls. As temperature rises increase, further effects will be felt on crops such as wheat, maize and rice that are vital to feed populations in fragile parts of the world. Rising prices are highly likely. The capacity for adaptation to changing weather patterns is always going to be difficult. Biggest challenges for farmers can be expected where the temperature rises are highest, and that is in the equatorial regions, where temperature rises may be in excess of 3 °C.'

Changing farming practices can improve carbon sequestration in soils. Finding new crops is a major consideration of land use to reduce emissions. Agricultural science has improved land management knowledge and practices. New methods for recycling biomass such as wood waste into building materials, landscaping products, animal bedding, paths and surfaces can reduce the need for landfill.

If we change our food consumption patterns, which can be accomplished by public awareness campaigns and pricing mechanisms, there is great potential to reduce greenhouse gases emissions from agriculture in addition to awareness of healthy food choices, and improved dietary choices can have net benefits in terms of improved human health, something that is clearly a priority. People can be encouraged to move away from over-processed, over-packaged foods that in many instances contain toxic substances, and are contributing to the pollution of rivers and oceans.

One of the key points for effective land use is the use of large land masses for bio-sequestration of carbon. Another key factor is to make it easier for agricultural activities to generate carbon credits that can be traded or sold. Agribusiness is therefore a key player, with a substantial part of the solution for global greenhouse gas reduction, by making land use more effective. Changing farming practices and better land use has a comparatively low carbon footprint and the potential to contribute massive reductions in the world's greenhouse emissions. Sustainable agribusiness has the potential to provide a clean, green engine of economic growth, given the international situation of rising food prices and land shortages. During the next few years of transformation of the industrial landscape into carbon neutrality, farming has a huge role to play.

Let's consider a couple of practical examples of agribusiness projects with the potential for carbon trading through emissions reduction, as well as economic and jobs growth from sustainable livelihoods. Mangrove regeneration and industrial hemp plantations are just two of many revolutionary agribusiness initiatives that are having multiple positive effects in the campaign against climate change.

With appropriate levels of financial investment, communities can be encouraged to adapt to climate change and sea level rise through co-management of mangroves and aquaculture. For a modest investment, these activities lead to the establishment of local jobs for sustainable fishing, food production, and dissemination of knowledge about regeneration of mangroves and rainforest that can be shared with other communities in similar regions.

One of the most vulnerable terrains to rising sea levels is tropical coastlines, where deforestation and removal of mangroves, vital fish breeding grounds, have occurred. Projects for restoration of mangroves and control of erosion and sea level rises, for example in Indonesia and Malaysia, have had an excellent effect on local ecosystems. Economic activities for local people include environmental management and generation of sustainable commercial agribusiness. While the existing conditions are understood in terms of stable past climates, more knowledge is required to manage the changes in the local ecosystems that have been caused by changing temperature ranges, weather, tidal and wave patterns, affecting the topsoil, coastal erosion and sedimentation.

Programs for the conservation and protection of coastal areas against climate change and global warming also increase the awareness of the importance of participating in the preservation of nature. Mangrove regeneration programs result in the creation of planting activities for the protection of coastal communities, through the establishment of greenbelt mangrove areas along the shoreline. As well as raising the level of awareness of the benefits of coastal protection, these programs mobilise local communities to actively participate in a wide variety of activities related to the protection and preservation of the environment, often with spinoff benefits of ecotourism business development.

Many local communities nurture the growing spirit of volunteerism by allowing people to engage with, and understand the importance of mangroves, and the role they have to play in having a positive impact on the protection and preservation of the coastal regions of vulnerable island nations. Island and shoreline communities, with a little financial assistance, can develop local jobs for sustainable fishing, food production, and disseminating knowledge about regeneration of mangroves and rainforest through ecotourism information centres.

Another example relevant to fragile countries is the development of an industrial hemp industry. Hemp is capable of enhanced carbon removal because of its properties, such as the ability to provide temporary cover between planting seasons, and cover bare paddocks with vegetation. This protects soil from the sun and allows it to hold more water and be more attractive to carbon capturing microbes. Plantations can also restore degraded land, which slows carbon release while returning the land to agriculture or other uses.

Although there is little in the way of comparative study of biomass growth rates amongst trees, plants and crops, hemp is one of the faster growing biomasses, producing up to 25 tonnes of dry matter per hectare per year. It can be produced organically, and its products are biodegradable. Hemp leaf is 50% nitrogen, enabling it to enrich rather than deplete soil. Historically, hemp was grown with crop rotation required only after a number of years.

It is perfectly suited to large-scale agribusiness, providing income to farmers, as well as photosynthesis of CO2 from the atmosphere. This produces organic compounds in the hemp crop, as well as in the soil in which the crop is grown. There is a growing international market for hemp building materials, hemp food products, as well as hemp textiles and fibre products. Canada, US and China markets have experienced substantial and rapid growth in production in recent years.

Hemp fibre is a crop that has been used since ancient times for food, textiles, and building materials. Once sown, it has unique properties for eradicating weeds, and stabilising soil erosion, helped by harvesting methods that allow the crop to provide ground cover for weeks after cutting, before baling. The fibre can contribute to the manufacture of bio-composite materials. The low density and highly crystalline cellulose content of the hemp natural fibre leads to excellent specific properties, which enable the use of this crop to compete with traditional glass fibres in structural applications without producing toxic residues.

Industrial hemp is not only a low carbon building, packaging and textile material, it also has the capacity to accelerate carbon sequestration in the soil, thus forming a natural carbon sink in land that could otherwise be responsible for increased emissions through soil imbalance. It offers some real environmental advantages, particularly with regard to the limited needs for herbicides and pesticides. It has a natural capability to be used as part of an organic agriculture strategy, and is part of a growing market for clean, green products.

The crop was outlawed in the US at the instigation of newspaper magnate Randolph Hurst, who feared competition for paper production. Germany has already developed industrial hemp non-woven products for the automotive industry. Canada has a policy that actively supports industrial hemp primary and secondary production, research and development. Australia has expertise in industrial hemp primary production research, as well as practical bio-composite research and development.

Hemp fibre has been subject to construction industry development and testing since the 1970s. Unlike tree fibre, where lignin is extracted from forestry products with chlorine, hemp building products do not create dioxins. Medium density fiberboard can be produced in the same machines that currently produce timber based products. As the fibres are much longer than fir or other timbers used in particle boards, hemp has the capacity to be used for structural building beams, because the strength of a product is directly proportional to fibre length.

The industrial hemp industry at this point in time is making efforts to consolidate and propagate specialist knowledge for the advancement of commercial viability for industry participants. As a food source in a world which is facing shortages and other food security issues, hemp is an easily assimilated source of protein which can be grown in a wide range of marginal conditions. Of course it was once a traditional staple in many countries until the exercise of political influence on governments that banned it because of a related plant with different properties.

BIO-COMPOSITES

Plant fibre bio-composites are gaining ground in replacing plastics and packaging materials. Bio-fibers and bio-resins can be integrated and eventually replace plastics, which are so damaging to the ocean, as islands of plastic debris are choking the waterways in every country.

Biodegradable plastics have an expanding range of potential applications, and overall, they are less harmful to the environment. However, some bioplastics are more 'environmentally friendly' than others. Recycling issues have emerged, requiring action to standardise recycling facilities for bioplastics. A major consideration is the rate of biodegradation and the disposal environment. Currently there is a range of biodegradable plastics available, some more practical than others for particular uses. International regulation is clearly required for bioplastics. This would provide the governance required to ensure quality and recycling methods.

There is an extensive range of potential applications,

such as plastic film, shopping bags, garbage bags, sanitary products, and bottles. The list goes on. The preferred method for bioplastics disposal is composting and soil burial. Many cities around the world now compost garden organic material, food waste, cardboard, and paper products. The recycling mechanism is primarily hydrolysis combined with aerobic and anaerobic microbial activity. As CO2 is a byproduct, so thought has to be given to effective disposal methods.

Municipal waste, organic waste, used industrial oils, separated sewage sludge, plastics, used vehicle tires and many other substances containing carbon and hydrogen can be used as input to new waste disposal technologies. These substances can be fed into equipment that produces gases containing carbon and hydrogen. A catalyst is used to break the existing carbon bonds in long chained polymers into single carbon fragments that can be recombined to form small chain hydrocarbons for recycled fuels, a process generally known as 'biofuels from waste'. There are some pros and cons, and as with bioplastics, attention has to be given to the energy used for processing. Some research and development is using quantum mechanical resonance to break the carbon bonds, saving on the large energy consumption from traditional methods of hydrocracking polymer bonds.

BUILDING THE FUTURE

The design of modular, self-assembly buildings and shelters for both urban environments, and remote and fragile landscapes is now a reality, with low cost, easy maintenance buildings and a minimal carbon footprint both in manufacture and construction. As well as improvement in design and materials, there are significant advances in rapid deployment of buildings with simple-to-assemble construction frameworks and building panels.

New sustainable, environmentally friendly materials are available for external and internal walls and roofing. Bio-resins and natural fibres are suitable for extruded stud frame houses and commercial constructions. The materials are not only low carbon in manufacture, but have demonstrably superior fire resistance.

In view of the risks of extreme weather events, and other emergency situations forecast to rise with the incidence of extreme weather events, it is a great advantage to have self-assembly building teams from local communities put up buildings such as schools, clinics, accommodation, site sheds, disaster relief shelters, ready to occupy in a matter of days. With current technology advances, this can include framework, plumbing, electrical wiring, and appliances.

Many bio-composite research organisations worldwide are committed to developing sustainable, environmentally friendly bio materials such as natural fibre reinforcements and bio-resins. This enables manufacture of frames for structured insulated panels, from bio-fiber reinforced polymer composites.

Self-assembly building systems can be developed for rural, isolated and developing communities. They can be constructed from natural fibre materials, to be lightweight, insulated, fire resistant, water resistant, termite and hurricane proof. And they can be assembled without the use of skilled tradesmen. Components can be manufactured using materials that have a lower embodied energy (energy used in production) than traditional construction materials. They are being designed as 'flat packs', for ready transport to building sites.

Assembly teams from local communities can put up buildings such as schools, clinics, emergency accommodation, site sheds, and disaster relief shelters within a short timeframe. This includes final fit out, as well as renewable energy and clean drinking water systems. Locally sourced natural fibre insulation such as industrial hemp, wool, or other fibres and materials, can increase the comfort and decrease the carbon footprint of the supply chain.

Community based developments have a particular need for robust simple technology that has been rigorously tested for use in difficult conditions. There is a global requirement for low cost, sustainable buildings particularly in view of risks from climate change. As well as local communities and households, government agencies and NGOs operating in difficult to access terrain and war zones can all be assisted by easy-to-assemble flat pack systems. New materials, systems and technologies are emerging to build rapid response shelters as well as more permanent structures. And the structures can be low cost, easy to transport, and easy to source, with collaboration amongst suppliers, research organisations, government agencies and communities.

LOW CARBON CEMENT

Portland cement usually originates from limestone. It is in commonest use around the world, the basic input into concrete, fibre boards, and mortars. It was developed in the 19th century, and its biggest disadvantage is that it contributes to more than 5% of the annual greenhouse gas emissions. This is because production of clinker as part of the manufacturing process, as well as the mining and the transport of materials, requires very high energy consumption. Emissions of greenhouse gases and particulates make Portland cements a clear target for replacement with more advanced cement based products.

Roman cements were naturally produced from limestones containing clay. First manufactured in ancient Rome, gypsum and lime were used as binders. Volcanic dusts were added, which made the concrete more resistant to salt water because of a high content of alumina and silica. This natural combination produced binders of great strength and durability. The success of the cement synthesis at low temperatures resulted from the mix of elements, silica, alumina and iron oxide not attained in any man made mixture.

Magnesium Oxide cement products, on the other hand, are a technology extension of traditional cements. They largely satisfy energy efficient criteria, in that, after initial firing, low energy MgO cement curing and production is mostly undertaken at room temperature, so that the embodied energy of the cement products is much less than any other type of cement. MgO cement has fewer process steps and requires a lower volume of raw materials. Prior to being turned into building boards, raw materials processing is undertaken at around 650°C, while production facility processes are mostly undertaken at room temperature. All of these factors account for significantly lower embodied energy from the lifecycle of manufacturing MgO building boards.

MgO cement products have been rated as achieving similar or superior strength, fire and water resistance and other mechanical characteristics and properties, at a lower volume, lighter weight, and with less input material than conventional fibre cement and plasterboard boards, with a longer life span and improved durability. They also exhibit increased biodegradability and recycling possibilities.

GLOBAL CARBON MARKET

The European Commission has estimated that 80% of carbon emissions in Europe result from electricity. Transport initiatives will see electricity and hydrogen powering the developed world's vehicles by 2020. For manufacturing, retail, mining, and the services industries, the major source of carbon emissions is still the consumption of electricity.

A clear target for carbon emissions reduction, such as substitution of renewable energy for fossil fuels, is the production of electricity. Carbon prices in the European Emissions Trading Scheme (ETS) are lower than expected because of, among other reasons, a lack of accuracy of the estimation techniques for emissions. Providing standards based ways for carbon pricing in real-time could stabilise and improve the carbon market.

Emissions Trading Schemes, as set out in the Kyoto Protocol, are reliant on market mechanisms of cap and trade to regulate carbon emissions. Different mechanisms are being used in different parts of the world (although probably the most developed is the European Union ETS). One Emissions Trading Unit (ETU) is equivalent to 1 tonne reduction of CO2 emissions.

There are a few emerging problems with cap and trade schemes, such as the estimation techniques having a significant uncertainty. Many emitters and industries are not covered by the schemes. Another factor has been carbon leakage. This refers to a net increase in carbon emissions because of different pricing and regulation between countries, of which carbon polluters take advantage. Currently there are no guarantees that these schemes will be effective in the actual measurement and reduction of greenhouse emissions into the atmosphere, because practically speaking, it is impossible to fully regulate a carbon market within the timescales required to be sure of limiting temperature rises to an acceptable range. Voluntary regulation has never worked particularly well in markets.

Initiatives for a clean energy super grid by Europe's North Sea countries (able to store power, effectively forming a giant renewable power station) were initially encouraging. With the lesson of the failure of self-regulation by financial markets still fresh, it is surely too risky to allow market forces alone to regulate carbon emission reduction. A better strategy for carbon emissions governance may be to selectively apply a regime of carbon emissions monitoring in near real-time to the electricity grid. Carbon pricing by generator source may serve to actively promote a popular culture of carbon emissions reduction, reflected in the price of electricity. This would provide requisite economic incentive for encouraging the production of renewable energy grids, and the development of micro grids. There are many possibilities to develop local area networks as well as wide area networks for power generation, using the best available sources of renewable energy. At the beginning of 2015, prices per kilowatt hour were dropping rapidly in response to the rapidly declining cost of producing solar energy. A price on carbon would seal the deal. This measure was not unilaterally adopted, resulting in a fragmented, incomplete and inefficient set of individual carbon markets, instead of a unilateral harmonised global carbon market.

Currently there are reasonably accurate metrics for greenhouse gas emissions from power generation, by generator type. By gathering data from transmission networks, a carbon price can be placed on real-time electricity generation according to the amount of fossil fuel generated power. Electricity supply is a zero sum game, and for traditional generators, supply has to meet demand, so that calculations made on the supply of electricity are an accurate source of information. They can be used by a real-time carbon market as the basis for pricing carbon discounts for energy sources that are not fossil fuel based.

Transmission Networks are often known as 'the Grid' or 'the National Grid'. They transfer electrical energy from electricity generators or power plants to electricity substations located near demand centres, often over high voltage networks. New electricity network technology, known as 'Smart Grid', combined with cloud computing is capable of integrating electricity measurement data with a web enabled trading market operating in wide area networks. This would provide spot prices for electricity, with accurate carbon pricing discounts for use of energy provided by renewable energy generation. Cloud data services can be used to aggregate power consumption, and associated carbon emissions, on an immediate basis, by generator source accessible from the internet.

By monitoring scheduled electricity transmission (and therefore consumption) by network operators, a mechanism of carbon emissions pricing discounts, applied to both industry and households, can be made transparent, and web infographics would encourage effective popular support. That is, offering cheaper electricity prices for greener consumption by application of a price on carbon.

A spot pricing market for electricity can thus provide a carbon price. Electricity transmitters, distributors and retailers can access a national energy exchange mechanism via a web portal, to buy and sell energy, receiving an associated discount for buying products with a lower carbon rating, effectively setting the carbon price. To achieve timely, accurate carbon emissions monitoring, it is important that a standard approach is used for the collection of electricity and trading market data.

Data about power transmission by generator type can be collected from Transmission Network Operators, both in real-time, and from forecast demand schedules. Data can be collected from scheduled transmissions, as well as from smart grid network devices that can supply near real-time metrics of consumption. As the take up of 'Smart Grid' technology increases, the data will increasingly represent real-time consumption.

Data gathering mechanisms do have a sufficient level of accuracy for market spot pricing, and the accuracy can only increase as electricity monitoring and demand forecasts improve. This provides an enormous opportunity to apply carbon emissions monitoring algorithms to demand and consumption data, providing carbon pricing discounts on the spot to wholesale and retail buyers of electricity.

From a greenhouse gas emissions reduction perspective, there is considerable urgency for the electricity marketplace to meet the increasing levels of renewable energy generation. Current mechanisms depend on having a small number of large generators that operate on dedicated power tie lines. This scenario is rapidly changing. In Europe, the European Commission has mandated that information technology is to be used to accommodate renewable energy generation, not the building of new infrastructure. A real time electricity market can offer customer discounts for use of renewable energy, creating further incentives for generators to move away from fossil fuel power generation. Discounts and consumption levels made easily accessible via the internet would prove very popular with consumers. Electricity micro hubs are starting to spring up at a local level. Integrating the existing networks with new micro networks is an effective way to promote the supply of renewable energy quickly and efficiently.

ENERGY EFFICIENCY

"Over the last twenty years using a Whole System Design (WSD) approach to identifying energy efficiency opportunities, engineers, industrial designers and architects have found they can achieve larger efficiency" - Commonwealth Scientific and Industry Research Organisation (CSIRO) Energy, Australia.

Energy efficiency can substantially reduce the greenhouse emissions from energy consumption for all organisations with large energy bills, from energy assessments. Initial site analysis can include building thermal performance, energy usage patterns, as well as the type of equipment installed. A review of energy provider peak and off-peak rates can be made to suit usage patterns. Detailed audits can be undertaken to gather data for input into the recommended energy efficiency solution measures. Smart metres and energy demand management can enable the fillip to further reduce energy bills and associated carbon emissions, by providing information about usage patterns and behaviours.

Carbon pricing would make organisations even more conscious of carbon footprints, as energy retail prices look set to rise further. In the retrofit of energy efficiency into buildings, as well as energy efficient technology, building thermal improvements can be applied to reduce energy consumption. Energy management by reduction strategies is a comprehensive approach to delivery of more savings than any other single mechanism, and of course is a lot cheaper.

Building thermal performance is a prime consideration when designing energy efficiency improvements. Replacement of lighting, energy usage monitoring, and installation of renewable energy can be designed with the characteristics of the whole site in mind, resulting in greater efficiency and proficiency throughout the process of energy consumption.

Understanding and predicting energy consumption behaviour to reduce electricity bills, also improves energy efficiency. Better products use less energy. For lighting, LEDs are not only the most efficient technology, the supply chain is much less toxic than fluorescent lighting. Customised design for particular buildings can produce very compelling energy efficiency results. Getting the right equipment for specific company sites can dramatically reduce consumption. While geo-exchange heating and cooling technologies use less electricity, no two buildings are the same. Specific improvements designed to suit the building's particular characteristics are the best way to maximise energy usage reduction.

Bill analysis of electricity consumption plays an important part in promoting efficient consumption. Scrutiny of historical billing data provides evidence of exactly what equipment usage patterns over time are costing the most. Change of consumption habits, tariffs, and equipment provides more savings. Smart meters, and the logging of usage data helps organisations to target the reduction of profligate behaviour. Best practice monitoring systems allow for observation of real-time data and comparative analysis from a web interface.

End-to-end design and a whole system view of complex buildings with multiple spaces helps reduce peak load by intelligent monitoring of energy usage. Examination of building characteristics and function can identify common sense improvements that further reduce the improvements from energy efficient equipment. Site specific renewable energy generation can provide income by supplying energy back into the grid. In many countries, there are a number of programs and funding options available for energy efficiency building retrofits. In some systems there are no upfront costs, with energy efficiency equipment paid for directly from electricity bill savings.

A site specific program plan can manage the delivery of energy efficiency improvements to building, plant and equipment. At this point in time the programs are mostly targeted at the commercial sector. Household consumption is a poor relative in the programs targeted at increased energy efficiency, and yet this is where the most pain is felt.

Smart meters monitor ongoing energy usage. Used judiciously they can significantly reduce carbon footprint. 'Advanced metering can enable businesses to identify energy, cost and carbon savings by providing detailed information about the way in which they use their energy. Although this technology is fairly well established in companies with significant energy demands, it is not widely used by small to medium sized enterprises (SMEs)'. Advanced Metering - Carbon Trust UK.

Access to energy data can provide metrics that help to optimise corporate energy consumption. Analysis by time of use, history, seasonal variations and peak/off peak patterns can make some low hanging fruit cost savings. Depending on the available meters, devices and sensors, and existing energy databases, near real-time data can provide significant cost savings from fine tuning of energy demand management and building problem determination.

Technology equipment savings can be analysed. Replacement of existing energy equipment can be an expensive business. Cost benefit analysis of equipment replacement can determine the best value for money, in view of the current costs of electricity. Site specific analysis can be applied to projected equipment replacement, based on representative sample data regarding the likely payback period, and the annual cost savings.

It is important to develop a strategic roadmap based on requirements gathering to deliver the full benefits of energy efficiency. Comprehensive pre-installation site specific surveys improve performance of energy savings from smart metering, lighting, heating, ventilation, and cooling. For some organisations the installation of wind, solar, tidal and other renewable energy generators can provide large cost savings over time. There is some excellent software available to collect data about energy consumption and production. Insights can be presented effectively in graphs, via reports, dashboards and, some programs provide KPIs of energy usage, and carbon offsets insights provided by near real time data. What can be measured can be managed.

At this point in time energy efficiency programs are mostly targeted at the commercial sector. Household consumption is a poor relative, and yet this is where the most pain is being felt as families struggle to pay rising retail costs.

RENEWABLE ENERGY

'Reinventing Fire' is the wonderful term coined by the Rocky Mountain Institute to describe the transformation of the pattern of energy consumption away from oil and coal, and nuclear energy. Careful analysis has to be the basis for finding new sources of energy, to ensure that the chosen renewable energy sources make the best use of local conditions to replace fossil fuels in the production of electricity.

So what is the problem? Oil, coal, and natural gas, the 'fossil' fuels formed by decomposition of dead organisms are causing excessive greenhouse gas emissions. All three were formed many hundreds of millions of years ago from the decomposition of plants and animals. Oil and natural gas were created from organisms that lived in the water and were buried under ocean or river sediments. Long after the great prehistoric seas and rivers vanished, heat, pressure and bacteria combined to compress the organic material. The processes of the carbon cycle that engage organic material are photosynthesis and metabolism.

During photosynthesis, plants use up carbon dioxide and produce oxygen. During metabolism oxygen is used and carbon dioxide is a product. This balance is part of the stability of the composition of the atmosphere, the air that we breathe. What happens when extra carbon that has been locked away in fossil fuel deposits is burnt and released into the atmosphere? The greenhouse blanket gets thicker, and the Earth gets warmer.

Nuclear fission, a chain reaction, releases the enormous energy stored in the nuclei of elements of matter. We humans have not learned to handle the tremendous forces unleashed by this process. It is an unreliable source of 'fire' because no solution has been found to its potentially lethal waste products, radioactive isotopes (such as Plutonium 239 which takes hundreds of thousands of years to lose radioactivity). Nuclear power is an accident waiting to happen, and the two major events at Chernobyl in Russia and Fukushima in Japan are just the tip of an iceberg, with nuclear plants located in regions that are geologically unstable. With the biosphere already in a state of uncertainty, pursuing nuclear power would be madness.

Nuclear fusion is one of the basic transformation forces of the universe. When nuclei are combined, matter is transformed, releasing massive amounts of energy. Einstein's famous equation $e=mc^2$ and some simple arithmetic demonstrates this fact. The speed of light means that photons of energy from the Sun travel the 150 million kilometres to Earth in about 8 minutes. Thus just a small amount of matter can produce a massive quantity of energy.

The good news is that our sun is our natural source of nuclear fusion. Deep in the core of our local star, hydrogen atoms react by nuclear fusion, producing a massive amount of energy that streams in all directions at the speed of light. And it is safely held at a safe distance from Earth by the gravity of spacetime, able to be harnessed by photovoltaics or solar cells.

Water is another predictable source of energy, with the cycle of precipitation causing our rivers to flow, and the moon providing tidal motion, both sources of mechanical energy that can be converted readily into hydroelectricity. One of the balance mechanisms of our climate is the winds that form from pressure gradients in the atmosphere providing consistent predictable currents of air, also providing mechanical energy that, like water, can be converted into electricity. It seems that as long as we take care of our climate cycle, we have abundant natural sources of energy that can be tapped with fairly simple, safe technologies, to meet all our energy needs.

The term 'renewable' means tapping into sources of energy that are natural, secure and have proven to be sustainable over time because they are part of the fundamental mechanics of the solar system that support water and carbon based life on our planet. Using a range of technologies suited to local environmental conditions, such as sunlight, wind, tides and rivers, there is abundant cheap energy available on the planet without having to use sources of power that fundamentally cost the Earth because of dangerous side effects

The availability, technology and cost effectiveness means that there is no reason that baseload power cannot be provided by 100% renewable energy. The caveat is that we need to use planning, analysis and foresight to ensure that the transition from our current monolithic generators to a combination of macro and micro grids is achieved as cleverly as possible. The biggest challenge is ensuring that the capacity of energy storage technologies match the renewable energy production.

Energy is a zero sum game. That is, energy supply has to exactly meet energy demand. Currently we are burning fossil fuels on demand. A more efficient way to meet our energy needs is to be able to harvest energy when it is available, for example the sun and the wind, and store it somewhere until it is required to produce power. Energy storage has to be part of the solution.

In January 2015, the Climate Group launched its RE100 program. It is a bold initiative that encourages energy companies to commit to 100% renewable energy. Their strategy and their analysis is excellent, however it does depend on investment not only in cost effective renewable energy generators, but also energy storage. Electricity can be converted into different forms of energy which can be stored and retrieved later to meet demand from consumers. It really depends on the length of time, how much electricity has to be stored, and what is available in the local area, as to which storage technologies are the most useful. A brief look at some of the best energy storage solutions that are key to enabling a 100% renewable energy grid shows that with investment, there are mature technologies ready, willing and able to replace fossil fuels. Even though there are associated capital costs to be met, the long term outlook is very profitable, and associated benefits are a sustainable renewable economy for us all.

For regions with a suitable topography, like Norway and Switzerland, hydroelectricity is a brilliant storage mechanism. The basic mechanism is to use excess energy to pump water up to a higher reservoir, from whence it falls to a lower reservoir, creating electricity by doing so. Large plants have high capital costs, and take a long time to design and build. Anywhere there are suitable natural reservoirs, enclosed to minimise evaporation are good places to establish solar, wind and other renewable energy generation capacity. The ability to use hydro storage means that stable baseload power can be delivered when needed.

Compressed Air Electricity Storage (CAES) also looks promising. It requires underground storage capability, such as artificial salt caverns. Current plants have installed capacity around 200 Megawatts. Thermal oil and molten salts are also being investigated as storage options. Hydrogen fuel cells are developing rapidly as the costs are decreasing. Utility scale hydrogen fuel cells can produce hydrogen with electrolysis, able to be stored for use in fuel cells to generate electricity. Supercapacitors are much faster than batteries and can deliver a lot of power in a very short period of time, so they are useful for fluctuations of power quality on the grid. There are many more renewable and sustainable storage solutions that are beginning to come into play, as investment increases, and we have the incentives in place to leave oil, coal and natural gas in the ground.

Micro grids have a significant role to play in providing access to clean energy. Large scale micro grid initiatives and projects currently underway in Asia and Africa. There is a staggering one and a half billion people who currently live in rural areas, most with no access to electricity, including rural Africa and the myriad islands in Indonesia and the Philippines. The forecast is for micro grids to supply around 40% of new rural capacity by 2030. Investment is key to developing this capacity. Currently there are a number of NGOs engaged to ensure commercial viability is being augmented by government policy, local skills and knowledge, and careful selection of renewable energy technology to match local resources.

Hybrid electricity generation based on local resources, and the potential for bioenergy as backup capacity, combined with new types of local storage solutions, may mean that micro grids may become mainstream in the short to medium term. By 2030, micro grids may become a reality for all communities, making today's national grids eventually obsolete. Electricity high voltage networks are inefficient distributors of power over long distances, with significant line losses, so it makes sense to replace fossil fuel centralised generation with multiple smaller renewable local solutions. A realistic price on carbon makes micro grids a good financial option.

Most people think of photovoltaic (PV) solar rooftop cells when they think of renewable energy. Today's technology is streets ahead of the first generation of solar panels. The photovoltaic effect is the use of sunlight to make use of semiconductor materials and the difference in their energy states to produce an electrical current. Nanostructured materials and photonic enhancements offer unprecedented opportunities to control both the optical and electrical properties of the next generation of solar cells. These cells can be produced at low cost, 3D printed onto building panels and public structures, for inbuilt electricity production. There is even a solar paint that can generate electricity from the sun. Before long, buildings can be provided sufficient solar capacity during construction for sustainable household use. Retrofitting could become as easy as a new coat of paint.

Concentrated Solar Power or (CSP) uses the principle of focusing energy in a relatively small area, by using mirrors and lenses to concentrate solar radiation which is converted to heat. CSP plants are capital intensive, but have virtually zero fuel costs. Parabolic trough plants however require thermal energy storage. Adding cost effective energy storage technology is the key to large scale adoption of CSP as a supplier of existing electricity grids. Operational and maintenance costs can be reduced significantly the more plants that are built.

Wave power makes use of dynamic ocean movements to capture energy perpendicular to the motion of the waves. Floats are partially submerged in the water, and when a wave rolls in, they are lifted upwards in succession by the wave crest. Wave motion generates hydraulic power which can be converted into into a steady power output to the electricity grid. Hydraulic cylinders supply power to a common fixed pressure system, which effectively is an energy storage system. This energy can be converted to electricity as required. Ocean wave power technologies are mature, and are in early stages of adoption into power grids.

Tidal power from rivers and seaways has some very interesting advances allowing for energy from current as slow as 2 knots. The vortex is a physical phenomenon that can manifest in water. Vortices can form in moving liquids and gases, a spiral that has properties of angular and linear momentum, and energy, as well as mass. A vortex turbine can use the vibration energy in the water current. The energy is a result of the difference in pressure between the core of the vortex, and the perimeter. The movement of the cylinder resulting from the pressure gradient is then converted to electricity.

Observation of a whirlwind gives an idea of how much power vortices can generate from relatively small, but constant input tidal flow. This form of energy has great potential for towns and villages near a water course where there is no access to an electricity grid. In the short term remote, developing and fragile regions are the prime benefactors of this type of technology, particularly in those regions where microgrid projects can supply most of the future energy needs.

Geothermal energy is an interesting option. It is heat from the Earth itself. It is clean and sustainable, and comes with its own storage. From shallow sources of hot water and rocks to molten magma, there is an enormous potential to exploit this untapped energy, with appropriate investment. Currently one of the early uses of geothermal energy is to provide cooling in summer and heating in winter for buildings. Geothermal heat pumps transfer heat into the indoor air delivery system. In the summer, the process is reversed, and the heat pump moves heat from the indoor air into the heat exchanger. This form of renewable energy has the potential to provide large scale grid power in geologically suitable locations.

Biogas is produced from raw materials such as agricultural waste, manure, municipal waste, plant material, sewage, green waste and food waste. As such, many local authorities are investigating and investing in biogas energy production, because it also solves part of the growing waste disposal problems that faces almost every municipality in every country.

TRANSPORT FUELS

The problem is to find practical alternatives, and transition strategies, so that petroleum combustion engines can be gradually replaced or upgraded to run on newer fuel types, such as hydrogen and electricity. The transition away from fossil fuels may be able to utilise the existing fuel stations as part of the renewable energy solution. Renewable energy technologies can now deliver baseload power (predictable, constant power in excess of 10 megawatts). A particular problem for the electricity transmission authorities, independent of generator source, is the ability to meet the demands of peak power supply, as power is a commodity that is difficult to store, and both households and industry are rapidly increasing their consumption of electricity.

Large scale solutions for storing energy are expensive, have a long lead time, and may not prove to be the optimal solutions to supply peak and off-peak power demands in a cost effective way. Smaller scale facilities can be more flexible, and more easily backed up to provide local power in case of electricity blackouts and brownouts. This also avoids the problem of a single point of failure affecting large geographic areas in case of extreme weather or even disasters. With the evolution of transport energy systems to electric motors, vehicles require new types of fueling stations to supply both electricity and hydrogen to fuel the new vehicles being delivered by manufacturers. A cost effective, energy efficient and low carbon solution may be to co-locate electric and hydrogen vehicle energy stations with renewable energy generators, using hydrogen and other forms of local storage to store excess energy for supply at peak times into the local electricity grid. Hydrogen may be particularly interesting, as it is not only a fuel, it can be used as an energy storage medium.

New renewable energy technologies provide a range of options, suitable for particular locations. Tidal power, wind turbines, solar-thermal and solar generators can all provide industrial strength energy at commercial levels. The conversion of petrol combustion engines to use electrical and hydrogen energy can be accelerated by the development of energy stations powered by renewable energies.

Hydrogen as a fuel for transport is now a mature technology. There are two main uses of hydrogen in vehicle engines. They are the adaptation of existing combustion engines to run on hydrogen, and the use of hydrogen fuel cells to provide power for electric vehicles. These fuel cells take in hydrogen and oxygen, and give off water, providing electricity to power the electric motors.

Hydrogen is also an efficient energy storage mechanism, and a two-way process can take electricity, store it as hydrogen, and then when required, provide fuel cells with energy that can be converted back into electricity, suitable for supply into the electricity grid. Electric powered vehicles, using either batteries or hydrogen cells as a power source, are becoming increasingly available. These vehicles require either electricity or hydrogen or both to operate. Co-location of vehicle depots with renewable energy production to supply electricity into a local electricity grid hub can provide a viable business model, particularly in the current situation of rising electricity prices.

The production of hybrid electric and hydrogen vehicles is slowly increasing. Acceleration of production of new energy fuels for vehicles is set to continue, with or without regulation of emissions, and associated carbon pricing offsets. Vehicle manufacturers such as Honda, Mercedes, BMW and Toyota are producing electric powered vehicles with hydrogen fuel cells for the marketplace, scheduled to make a serious entry in 2015. Clearly new vehicle fuels require new energy resourcing stations, however in the early periods of operation, this is unlikely to be profitable until the take up of the new vehicles becomes widespread. The cost of production of fossil fuels is rising as supply is declining. For enabling vehicle fleets with clean energy, the problem remains to provide convenient supplies of hydrogen and electricity, particularly in view of rising electricity prices because of the need to bolster ageing grid infrastructure to meet increasing demand. New energy supplies for vehicles can facilitate the transformation of land transport fuel consumption. The technology also provides improved engine performance, particularly with the advances in hydrogen fuel and fuel cells.

Many companies are engaged in the development of vehicle engines, and as the technology matures, costs are coming down. In January 2015, Toyota offered royalty free use of the fuel cell related patents related to fuel cell stacks, high pressure hydrogen tanks, fuel cell system software control, hydrogen production and supply. This is clearly going to be a game changer, and by 2020 hydrogen stations for transport may well replace service stations selling fossil fuels such as petroleum, diesel and even ethanol. (The problem with biofuels is that very often they are being grown in place of food crops, in regions where agricultural land of food is in short supply, and is not biosphere sustainable.) The collocation of renewable energy and hydrogen stations for vehicles may yet prove a winning combination.

COP21 PARIS

Up until now, we have placed no monetary value on either unpaid labour, much of it performed by women, and on the natural resources of freshwater, forests, clean air, healthy soils, and unpolluted rivers, seas and oceans. This is clearly a situation that has to be addressed as part of the measures to stabilise the biosphere, and has to be implicit in the activities that are funded to address the reduction of greenhouse gas emissions. Putting a value on clean water and non-toxic food supply is surely a starting point. Funding community based enterprises into sustainable agribusiness, environment management, and micro businesses for renewable energy is a priority of the United Nations Environment and Development programs, and there are many links between these programs and the UN Framework Convention for Climate Change. Encouraging the leadership and the full participation of women in these initiatives is seen as a very effective strategy for economic development in fragile regions.

The United States, the European Union and China, which together account for half of all greenhouse gas emissions, have pledged to do more, however it may not be

enough. All nations are expected to make their emissions cutting commitments known in the first half of 2015, and these pledges will send important signals about the ambition of the final agreement to be reached in December 2015 at the 21st Conference of Parties in Paris.
Mandatory carbon cap and trade schemes worldwide would see a rise in carbon price. Currently the low price means that the largest emitters of greenhouse emissions do not spend as many mandatory carbon credits on reducing their own emissions, because it is cheaper to meet commitments with projects that are successful, but do not have a very significant effect on overall reduction of emissions.

A significant rise in the carbon price would mean that the large polluters would address their own emissions, rather than spending their emissions reduction budget in less developed parts of the world. It has been very difficult to get Emissions Trading Schemes up and running, as countries put carbon pricing jurisdictions in place. An ETS guarantees the outcome, but not the price. Taxation, on the other hand, guarantees the price but not the outcome. With an increasing concern for meeting United Nations targets for reducing emissions to keep global temperature rises within defined limits, more countries are supporting an outcomes based approach.

The EU trading system was the vanguard, and other jurisdictions have been able to learn from the early experiences. Increasingly each year, regional carbon markets are growing. They had a value of 30 billion USD in 2014. Multinational corporations can clearly benefit by an integration of the separate carbon markets into an overarching international system. China is on track for a national trading system during 2016, and a gradual harmonisation of local systems is a stated goal. The United States EPS has new rules for restricting emissions from existing power plants. The California cap and trade system is being considered by other states. Many countries have national and regional trading systems.

Linking of international Emissions Trading Systems has proved challenging. When the building blocks of the system are identical, there is no real difficulty. However, most systems have technical and structural differences, as well as political differences. While there is a real intention to harmonise regional markets into a national system, the problems and challenges of integrating data and information that are not standard have become obstacle courses.

This is not a new challenge; in fact, it has been faced by other industries trying to make diverse systems communicate in real time. Notable examples are banking to connect live financial markets, and the utilities initiative 'Smart Grid' for balancing supply and demand across regional borders. The Information and Communications Technology industry has been at the forefront of meeting the technical challenges of facilitating information from different sources to be standardised, organised, distributed and managed from heterogeneous systems. Experience shows that taxation systems can coexist with a market pricing scheme. Taxation can be used to underpin a low carbon price, although this leads to a discrepancy in the cost of emissions reduction. The adjustment of taxation with offsets is similar to the way an emission trading system works, and this means that adjustments can be applied on a regional basis. Taxation can be used for protection of national industries, and for some countries this may be an attractive option.

A multiplicity of systems means that currently there is a huge range of prices for emissions in a global sense. In 2014 this was a discrepancy of more than 250%. Clearly this does not guarantee the outcomes required, and that is a focused, consistent reduction in emissions targeted at the industries that are the major culprits for greenhouse emissions, and thus global warming and climate change. Perhaps the lesson to be learnt is that targeting the major sources of greenhouse emissions at source as a starting point for an international Emissions Trading Scheme is essential to guarantee the results. Chinese experience shows that voluntary engagement in carbon trading only works to a small degree.

In the international context, it is clear that the wealthy corporations who are the major polluters continue to be self-serving, meaning that further policy and market mechanisms, regulation and public pressure all have to be applied to build up a comprehensive collaborative approach for real reductions of greenhouse emissions. We have to encourage, manage, and police the emissions reduction targets so that they meet the objectives set out by the scientific bodies charged with advising the countries engaged in the Kyoto Protocol negotiations. Near enough is not good enough when it comes to meeting the clear scenarios identified by the Intergovernmental Panel on Climate Change for limiting temperature rises to as close to less than two degrees Celsius as possible. Most organisations now recognize that we have missed the boat to achieve the lowest global temperature band rises, by failing to achieve concerted action to date. By 2011, we had already emitted about two thirds of the maximum cumulative amount of carbon dioxide that we can emit if we are to have a better than two thirds chance of meeting the 2°C target.

Because of our past performance burning fossil fuels for electricity, transport fuels, and the manufacture of concrete, even if emissions are stopped immediately, temperatures are going to remain elevated for centuries due to the effect of greenhouse gases. Limiting temperature rise will require substantial and sustained reductions of greenhouse gas emissions. Financial resources are moving into climate adaptation and mitigation on an increasing scale in the lead up to the new international climate agreement to be forged in Paris in 2015. There is a window of opportunity to take concerted, cost effective action to reduce emissions. However, it is a brief window.

New directions are required for adaptation and mitigation of climate change. Positive action and enhancement to current financing processes has to take place. We cannot fit sustainable technology projects to set guidelines, as we do not yet have the depth of knowledge or experience to do so. The approach has to be to fit the governance to the projects. We must adapt. The winds of change for financing developing country climate change projects are blowing strongly. The UN FCCC Scaling Up Climate Finance participants seem united on that count. There have to be new pathways for mobilising climate finance, to be agreed and instituted at the UN Conference of Parties COP 21 in Paris in December 2015. There also have to be pathways to community project finance in the developed world, as this is where innovation is most likely, and where the communication networks to communicate new and improved technology are the most powerful.

GLOBAL KNOWLEDGE SHARING

Greenhouse gas emissions reduction can be summarised by just a few activities. Firstly, replacing fossil fuels with renewable energy is key to reduction strategies. Sequestering carbon directly can be achieved by planned stabilisation of the natural environment such as finding alternative activities to logging and overfishing the oceans. Sequestering carbon indirectly can be achieved by planting crops that have high absorption properties, and the use of building materials that absorb greenhouse gases. Making communities energy efficient and sufficient through public awareness campaigns to change the way people and organisations use energy is also a highly effective and critical measure.

Anecdotal evidence indicates that many community projects capable of reducing emissions are foundering on the rocks of slow and inefficient climate finance. It is not only habitat that has been sacrificed on the altar of financial gain, investors and governments have been loath to fund greenhouse gas reduction research and technology. For example, market-ready renewable energy, bio-composite materials to replace plastics, carbon sequestering agribusiness, energy efficient construction technologies and energy from waste technologies have been unable to gain funding. New pathways are required, as well as lightweight governance mechanisms for mobilising climate finance to accelerate greenhouse gas emissions reduction. Micro and direct finance can provide a kick start to climate mitigation projects and clean technology innovation.

Greenwashing has to be curbed. It is imperative to provide robust processes engaging relevant climate professional organisations and individuals, to ensure that money is effectively spent. The pooling of knowledge and experience about climate finance and science can expedite emissions reductions, by making finance accessible to small projects in both developing and developed countries. Climate change does not respect borders, and urgent action and resources are required for projects in every corner of the globe. The expertise, technology, people and organisations are already in place.

It is time for an international initiative to fund the projects that are ready to go, with technology that has already been developed. Why continue to try to fit sustainable technology projects to inadequate guidelines that can be politically exploited or easily corrupted? The approach has to be to fit governance from appropriate professional scrutiny directly to the projects.

Funding channels and instruments can be applied to local community projects and innovative technology. Professional evaluation, can provide simple online pathways for approval to mobilise climate finance. Web workflows can be backed up by specialist knowledge, experience, case studies and skill sets to enormously speed up climate finance delivery.

A public search facility can ensure that scientific and economic analysis can be recorded and made accessible to get the right information to the right people at the right time in support of climate funding approvals processes. Simple intuitive interfaces can facilitate the uploading of project data, photos and text, subsequently linked to geospatial and scientific information, analysis and other data. Mapping of data can be automated in the background.

A web submissions process can enable real-time feedback for rapid clarification of information, ensuring that project approvers gain a real sense of the value of a project to the local community and economy, as well as a reasonably accurate estimate of the associated emissions reduction. International expertise and knowledge can be readily accessed to solve local problems. Cloud hosted reporting by project, ecology, climate systems, as well as spatial location, can provide insight into emissions reduction measures, local ecologies and micro climates. This information can then be shared globally and publicly.

Climate finance and scientific expertise can engage in capacity building and transfer of technology, ensuring that existing information can be shared to enable accurate assessments of climate change risk to communities. Information technology monitoring can place a real value on the natural environment to the global economy.

CLIMATE RISK MANAGEMENT

Disaster response in view of recent extreme weather events, such as hurricane Haiyan in the Philippines in 2013, and the increasing annual devastating bushfires and floods in Australia requires additional effort to mobilise local and international large scale relief efforts. Collaboration of organisations and processes has to be a priority. It is apparent that there is a pressing need to make the right information available at the right time to enable flexibility for gathering essential resources, such as water, food and medical aid. Access to knowledge and existing information is critical to ensure that responses make the best use of available resources. And yet immediate crisis response is only part of the story.

Risk management begins with mitigation. Climate mitigation and adaptation activities and projects have to be financed on an increasing scale. The recent IPCC AR5 report found that not only is Earth's climate warming, the rate has increased over the past few decades. The ocean is heating up, storing more carbon, and becoming more acidic, so that animals' protective shells are weakening. Globally, regions of ice and snow are decreasing. The number of detailed scientific reports, anecdotal evidence from local knowledge, findings and observations is accelerating to the point of information overload.

And yet collection, storage and ready access to information can be mobilised for global access. Climate risk related information access can be automated for public and private purposes. Decisions can be informed by interactive, location based mobile applications. Instead of multiple uncoordinated efforts, consolidation of pathways to climate data can be orchestrated to ensure effective use of resources, people, finance and information. Real-time text translation of conversations, data and information can promote improved response to the growing number of challenges, by facilitating collaboration amongst all stakeholder groups engaged in climate related activities. Particularly important is managing responses to disasters, and events that may lead to loss of life and damage to property and community.

Technology can provide the basis for a coordinated global response to the diverse activities of people and stakeholder organisations engaged in addressing climate disaster management. Messaging can not only facilitate collaboration amongst people and organisations, it can be harvested after the event for response analysis. Mobile data and the internet can provide timely information. The prerequisite is to establish a common terminology and a translation facility for search, access and display of information resources relating to managing risks and disasters stemming from extreme weather events. A common pictorial taxonomy would be a key asset.

There is also a clear requirement to provide access to scientific data from a search capability. Facilitation of communication amongst peers is going to ensure that data is interpreted consistently and accurately. Taxonomy mapping can provide data across scientific disciplines. Flexible response mechanisms have to be devised so that valuable knowledge can be shared internationally. Risk as it relates to the exposure and vulnerability of communities to climate extremes, has to be addressed with ready, easy-to-use communications.

Currently, data on disasters and disaster risk reduction are lacking at the local level, inhibiting improvements in local risk reduction and rapid response. Local knowledge databases can help mobilise disaster relief. Appropriate sharing of information across communities with similar problems can also improve the coordination of individual efforts. The key is providing data access when it is needed. Humanitarian relief is going to be called upon more often if prevention measures are inadequate. Logistics for the supply of relief and reconstruction resources can be further automated and communicated to improve response times.

Data collection mechanisms for resources for risk

reduction, disaster relief and reconstruction projects can be developed. Communication and collaboration with peer groups is essential for mobilising the best possible preventive measures. Review of best practice can be shared globally to ensure that mistakes are avoided. Timely specialist advice can provide better decision making. Sharing information about disaster risk management and climate change adaptation into local, regional, national, and international development policies and practices can provide many benefits. Relevant information about methods and practices for responses to threats and risks can be made easy to search. Practical information and specialist expertise can be provided on demand.

Opportunities that exist to create synergies in international finance for disaster risk management and adaptation to climate change have just begun to be explored. One way to identify these synergies for disaster risk management, mitigation and adaptation to climate change can be facilitated by simple, pictorial web pathways to relevant information for common scenarios. Making information easily accessible can prevent the reinvention of wheels for disaster response and risk mitigation. Experience already shows that stronger efforts at the international level do not necessarily lead to effective, rapid results in local communities.

Local knowledge can be presented in context with scientific and economic data to improve the quality of response to climate change. Harvesting local knowledge is critical. A combination of scientific, technical and economic information resources can help communities to develop realistic solutions on the ground. Sharing of subject matter expertise, problem solving in similar situations is an important currency for propagating best practice. Encouraging people to share information using internet communications is a very effective way to reduce risk. Being able to access knowledge and information automatically in the context of geographic location can make the difference between life and death.

And of course, for disaster relief and reconstruction, it is essential that communications are established to enable access to knowledge and expertise exactly when the need arises. These functions are well within the capability of current computing technology systems.

CLIMATE FINANCE PATHWAYS

For data exchange about climate projects, science and economics, standard information taxonomies are absolutely essential. They are the first step to international public information shared by geographic location. The deployment of global climate information systems can provide dissemination of local knowledge from hands-on experience, as well as established climate science.

Most countries are responding to climate change with specific activities addressing agriculture, horticulture, silviculture, and aquaculture as well as improvements to local ecosystem management. All industries can benefit from sharing knowledge about non-proprietary research and development.

There is a unique opportunity to provide real-time location connected data for a wide range of climate mitigation, adaptation and remediation activities, fostering collaboration and knowledge-sharing of solutions for particular problems with respect to maintaining human habitat while enhancing regional biodiversity.

A searchable, publicly available, common climate information service can ensure application of climate science, and economic data in the context of new community and regional climate projects, by ensuring that local and global knowledge is widely, publicly and rapidly accessible. A standard cloud technology approach can accelerate information dissemination to people and organisations involved in addressing climate adaptation, mitigation, risk and disaster management

Human timescale (rapid response) data exchange depends on using a combination of high speed search technology indexed by a common information model. This is an efficient and effective way to deliver meaningful global climate information from distributed data centres via mobile phones and internet to local communities.

Trust and confidence, open conversation on the strategies for adaptation is required. Donor countries have different national systems for accounting. We have to streamline accountability. Better, lightweight governance can have positive effects, as climate finance money becomes less scarce. Micro finance can encourage the sharing of roles between donors and developing countries. There is no single answer for the enabling pathways to climate finance. Flexibility and agility have to be the watchwords to allow human beings to do what we have been doing for several hundred thousand years, and that is to adapt to change.

We know that there is a correlation between the

levels of greenhouse gases in Earth's atmosphere and temperature, and that at a certain point climate feedback loops can tip the balance in life supporting ecosystems. The clear pathways to support slowing the rate of CO2 emissions are to increase renewable energy, stop logging old growth forests, monitor emissions from supply chains, and to make buildings and habitats energy efficient. The good news is that there is plenty of expertise, case studies and metrics in all of these fields to develop new sustainability technology projects that could be funded into the community by online project financing services.

This is largely true for both developed and developing world countries. The biosphere does not have national borders, all of the earth's micro climates are inextricably linked to the whole planetary ecology. Rather than just focusing on national approaches to climate policy, and individual projects with limited competencies, the next step has to be large-scale enabling of projects that can pass the science scrutiny. Governance mechanisms can be advised by climate scientists and sustainable technology organisations, who can be empowered to direct projects that result in rapid and sustainable greenhouse gas emissions reduction, not only in the developing world, but also in wealthy donor countries, where the politics are also infiltrated with vested interest.

Everyone pretty much agreed on the outcomes of the low carbon economy, where clean, green sustainable business provides social and fiscal benefits for the many. There is room for multiple approaches to ensuring real projects with real merit in reducing greenhouse gas emissions can get off the ground, rather than the situation commented upon by one national delegate to the September 2013 UN Climate Finance Pathways meeting in Incheon, Republic of Korea 'We create so many plans, frameworks and policies, that we do not have the capacity to implement real solutions'. Positive action and enhancement to current finance processes has to take place. We cannot fit sustainable technology projects to set guidelines, as we do not yet have the necessary knowledge and experience set in stone. The approach has to be to fit the governance to the projects.

Climate related projects can take into account local emissions factors and select the right projects to provide real emissions reduction. Sharing of knowledge about climate ecosystems has to be uninhibited by geography and national borders. The funding of projects, not only within country ownership and national fiscal frameworks, has to allow individual projects and local eco groups to apply for climate finance directly, free of national agendas. Prevention of greenwashing can be served by developing a network of ecology, climate, technology professionals and organisations to serve on governance panels for particular projects within their sphere of expertise.

We have to abandon the approach of applying traditional economic indicators to climate sustainable technology (which do not work, as the natural resources of the environment are not currently assigned an economic value), and instead focus on reduction of greenhouse gas emissions as the key success factor as a measured and measurable result, preferably linked into an international price on carbon. So-called 'donor' communities may be looking for traditional metrics, based on statistical series collected over time, to provide evidence of strategic value. A more agile process is required, given the urgency indicated by the overwhelming evidence collected by climate science. We have to lose the mentality of a developed and developing world finance patronage. We are all in the same climate change boat.

Above all we have to develop the habit of rewarding innovative methods of doing business, as it is only lateral thinking and new approaches that can save us from short term thinking, the kind of thinking that caused the problems in the first place. Is the UNFCCC framework too cumbersome? We have to change the framework! Flexible approaches are needed. Country based priorities are not enough, we require strategic global approaches as well. One of the human shortcomings of any program in any country is that agendas can be skewed by vested interests.

Economic transformation can accompany adaptation and mitigation strategy. Direct access by communities to finance has to accompany current investment mechanisms. The most important enabler is a strong policy signal from leaders to take action, and that a price on carbon takes the impetus away from donor countries directing action, usually from an incomplete picture of the problems, based on political concerns, rather than a global view of the health of the planet's biosphere.

Lessons can be learnt to apply to the climate finance agenda. Most countries have highlighted the importance of direct access to climate finance. Aid effectiveness analysis does not necessarily lead to better outcomes. Policy frameworks focus on sustainability, and this is a field that is qualitative. Greenhouse gas emissions are quantitative, not political, and the clock is ticking away on the energy balance of the biosphere on our blue planet.

Social media, and the setting up of better communications between climate scientists and climate finance people can be a very powerful tool to ensure that news, views and experiences on the ground can be exchanged between climate sustainability projects and financiers. The human race has survived extreme climate events over several hundred thousand years. We did not survive because of protocols, rules and regulations. We survived because of our innate ability to develop technology to adapt to changing conditions.

TAKING RESPONSIBILITY

In January 2015, one of the traditional newspapers in the UK published a headline of climate scientists begging governments to ensure that the rest of the fossil fuels, natural sinks for greenhouse gases, remain in the ground, sequestering rather than releasing them into the atmosphere. The old saying that we the people get the government we deserve seems apt. Do we change governments, reflecting a healthy attitude to our own future, our children's children, and this amazing planet that has nurtured us for hundreds of millennia? To reduce carbon emissions from burning fossil fuels and deforestation of old growth forests is the choice we have to make, and we don't have long to make it.

This work started with a historically attributable quote by Gregoire de Tours on the collapse of a mountain in the 6th century, when the local people had sixty days warning in the form of loud rumbling noises of the instability of the rocky terrain. And yet a large number of them chose to ignore the signs, and were observed to perish in a terrible way, as the walls of rock collapsed on their settlement, or were washed away by the torrential rising of the river to terrifying heights. It is not the knowledge and technology that are lacking, it is the collective will of the people to empower ourselves to make the requisite changes, to acknowledge the reality, understand the consequences of failing to address climate change. Now is the time to act promptly and decisively to avert impending disasters by understanding that it is the symbiosis of Earth's biosphere that currently nourishes and supports our wellbeing. Time to stop the denial of the havoc we are wreaking on this planet by failing to acknowledge the value of the natural world. Time to honour the ancestors and the incredible sacrifices they made so we can have the life we have today. And that includes all the plants and animals as well as us human beings.

There are clearly some human behavioural anti-patterns that are inhibiting the determination and speed at which we address climate change. We accept bad behaviour by wealthy international corporations, as though they are just a fact of life. Of course, we are the shareholders. So the question becomes, how do we achieve the critical mass of public consciousness needed to change the way we deal with the urgency of limiting greenhouse gas emissions? From the experience of the ancestors, and the history of human achievements to date, it is clear that it is neither conquest, nor adhering to tradition, nor by being wealthy, nor by being wise after the fact that can address the causes of runaway climate and weather variations.

There is only one primary attribute of human behaviour that has been consistently successful in ensuring our survival, and that is our ability to thrive under changing environmental conditions. This has been demonstrated consistently over the period that we have managed to make a record of our collective past, based on DNA analysis of migrations, archaeological evidence and climate records as well as written anecdotal accounts. It is our ability to innovate with technology, to think our way out of a crisis, and to act decisively and determinedly to change past patterns that no longer work for us that are going to count from now on.

This is a vital time for humans to evolve to the next stage of consciousness. Time to learn to live in harmony and peace with our natural environment, maintain the remaining wilderness areas, improve the sustainability of our food production in consideration of the long term health of the land, and above all, to respect the other animals as having the same rights to exist as we accord ourselves. It is high time to abandon the old economy of short term profitability with no value placed on the natural environment. Can we adopt and adapt to our collective new role of custodians of this beautiful blue planet, the only home we know, in time to stabilise the atmospheric conditions and the oceans, the forests and the soils, before living systems change so quickly, in such unexpected ways that we humans have no time to adapt? It has happened in the past. We humans can change our ways. We can avoid becoming one of the casualties of the sixth mass extinction event on Planet Earth.

Throughout the millennia, information, knowledge and education and an innate passion for technology innovation have been the major evolutionary influences. Counteracting this has been unhelpful behavioural patterns that have had a devastating impact on the human condition and the biosphere, the living systems of the planet. Life on Earth has its own destiny, it is not dependent on human beings, and like the sorcerer's apprentice, we meddle with the planet's homeostasis at our peril.

When the solar system formed, although the level of our sun's radiation was less than today, the early Planet Earth was warmer. It turns out that the level of greenhouse gases in the atmosphere has been self-regulating for 4.5 billion years, keeping the temperature range of the planet capable of supporting the evolution of stable carbon based life. As the sun gets brighter and hotter, atmospheric CO2 is deposited as carbon into the rocks. This causes the level of greenhouse gases blanketing the planet to decline, cooling the planet. The carbon cycle has been keeping the Earth's temperature stable, despite the increased level of radiation. The current situation is that human activity has been pumping additional CO2 into the atmosphere, causing a corresponding increase in the warming effect of the sun's radiation trapped by the layer of greenhouse gases. This is the effect that climate scientists have discovered is changing the planet's natural carbon cycle.

The carbon cycle is a dynamic process; however, it is not clockwork. As far as optimal conditions for life were concerned, the pendulum has swung too far in the past for a healthy biosphere. Twice in the last two and a half billion years, before humans evolved, the planet became a giant snowball. As the tectonic plates shifted to the warmth of the tropics, it caused a faster rate of rock weathering, which in turn produced chemical reactions that ended up locking more carbon into limestone. Snowball Earth conditions were eventually stabilised because of volcanic activity, although it took a long time, because ice cover is a powerful amplification mechanism, more ice reflecting more heat back into space. It shows that feedback loops keeping the temperature range of the planet stable can cause extreme effects. The Earth is orbiting the sun at a temperature range where water is present as a liquid, providing a habitable zone for carbon based life. Other planets, such as Venus and Mars may once have contained water, but today are too hot and too cold respectively to support life as we know it. Earth's climate has lurched from one extreme to another several times since the planet was born.

The climate has been changing from cold to warm and back for millions of years. How does this affect us humans? After all, we have only been around for a couple of hundred thousand years. People have been adapting to natural variability in the climate, however humanity only just survived the last glaciation, with the population numbers declining dramatically. Predictable changes take place caused by differences in Earth's inclination and orbit. In the past hundred years, increases in temperature cannot be explained by these effects. The rise in the level of atmospheric CO2 is the cause, triggered by burning the deposits of fossil fuels, vast quantities of which were ironically laid down in one of the great mass extinction events. Petroleum and natural gas were formed by the anaerobic decomposition of organisms including phytoplankton and zooplankton that settled to the sea bottom in large quantities when the oceans were starved of oxygen. Most deposits were a result of the Permian Triassic mass extinction, over 250 million years ago, nicknamed The Great Dying, since a staggering 96% of species died out. All life on Earth today, including us human beings, is descended from the 4% of species that survived. The exact cause is not known, though suggestions range from massive volcanic activity to a runaway greenhouse effect triggered by sudden release of methane from the sea floor.

Geophysical observations of sediment in lakes and the oceans, and analysis of wind and wave patterns provide rich sources of data for analysis of past climate patterns. Past temperatures and rainfall can be read from annual tree growth. Volcanoes cause climate variation, and in the short term, ash in the atmosphere reflects radiation back into space. Natural solar variations also cause increases and decreases, and this has occurred several times in recorded history. Ice core data, from the Antarctic and Greenland ice sheets and glaciers provide time series analysis for air bubbles trapped in the ice, providing data from as long ago as 800,000 years.

The loss of ice mass in the Arctic is a function of the rise in ocean temperatures, melting glaciers and calving of icebergs at the periphery of the ice sheets. While Antarctica currently has no net loss, its stability is threatened by wind and sea temperature changes. Ice core data gathered from annual ice forms (rather like tree rings) at the Vostok base in Antarctica has provided valuable information of past temperature variations and their correlation to the amount of CO2 in the atmosphere. The level of deuterium (heavy hydrogen) measurements is a function of, and can be used to determine air temperature variations. The interpretation of Vostok ice core data, over a long period, shows a close correlation between Antarctic temperature and atmospheric concentrations of CO2. The changes are associated with glacial to interglacial transitions.

When the Earth was born, the atmosphere was predominantly composed of carbon dioxide. The past trend however has been variable with an overall steady decline. In the geologically recent past, particularly since the evolution of humans, the atmospheric concentrations of CO2 have been relatively stable. There are seasonal variations, because of plant life cycles, and regional variations, due to the large metropolitan areas with high levels of traffic and industrial activities. Past records show atmospheric concentrations of CO2 ranging from 180 to around 300 parts per million. However, since the industrial revolution, levels have been rising steadily. In 2014, levels of CO2 rose above 400 parts per million for the first time, unprecedented during the past 800,000 years.

Global temperature rises dance in synchronicity with the level of atmospheric concentration of emissions of greenhouse gases that cause a blanketing effect on the Earth, warming up the surface. While water vapour is the major greenhouse gas, the addition of extra carbon dioxide by our industrial activities has provided the lion's share of global average rises in the past hundred years. Because it is unknown how and when feedback loops, such as melting ice sheets and changing wind patterns, are going to interact with one another, we may well be on very shaky ground. Presently we cannot accurately predict the complex effects between the climate systems of Earth's biosphere, including global temperatures. The result? We just do not know exactly how the systems that make up our climate are going to respond as a result of the warming in the oceans, on the land and in the atmosphere. Because although we can measure some of the results of our experimentation with burning vast quantities of fossil fuels (thus releasing correspondingly large quantities of CO2), we simply cannot predict the way feedback loops are going to change the climate systems and thus the weather. Nor do we understand all the causes and effects that we have unleashed. The inference is that we have a high probability of letting ourselves in for a period of certain climate instability over the next millennium.

The time to act is now to gather the information required

and to develop technology to address the risks of further destabilisation to the natural environment and human society. And the imperative? We know that climate change is accelerated by emissions of gases such as carbon dioxide and methane, so we simply have to stop burning fossil fuels. This means continuing to upscale the renewable energy sources for the major emitting countries, China, the US, India and Russia. (It is worth noting that Australia is the largest source of emissions per capita). Most of the emissions are caused by the burning of fossil fuels for electricity. The other notable major source of emissions is deforestation. A global carbon market is urgently required.

Time to get serious about addressing climate change, before drought, flood, fire and famine, cause more havoc than capacity for risk and emergency response than we can handle, irrespective of geographic location. Collectively, we have to use all our innovative skills and knowledge to find a way to stabilise our weather systems, and deal with the fact of unstoppable rising sea levels, even as some nations are already slowly submerging beneath the waves in the Pacific. All responsible parties and governments agree, unfettered emissions from burning fossil fuels are completely unacceptable. We have to address the root causes of the risks to the planet's stability, the level of greenhouse gases in the atmosphere, blanketing the planet. We have to respond to the sudden onset of global temperature rises and the melting of the ice sheets and glaciers, before the only response left to us is disaster management.

Time to examine whether we can use our current technologies to address the inherent causes of the problems of the accelerated global temperature rises. We clearly have to address our social and cultural issues, as well as the technological and scientific solutions. And we have to put a price on carbon, and a value on the natural environment.

Time to commit deep financial and human resources, to find the best possible solutions for reducing emissions. We do have to be careful to avoid the unintended consequences of early adoption of new technologies by providing ready access to shared knowledge and information. Our very existence is under threat from global warming and climate change, symptoms of neglect for the biosphere that sustains us. We have to collaborate to survive.

PART 3 - OUT OF THE PAST

THE ORMONTS VALLEY

Dense forests were able to grow in Switzerland only since the last great melting of the ice sheets that covered the mountains, reaching from the snowline all the way along La Grande Eau, the stream flowing rapidly over the moraine from one end of the valley to the other. The colonisation of the Ormonts Valley may be shrouded in mystery, however over the millennium and a half since the Tauredunum event, as for everywhere else on the planet, human activities have slowly changed the landscape in the Ormonts Valley.

The character of the valley is determined by the geological structure. Ormonts is composed of sandstone and shale, easily weathered, allowing for more cultivated land on the gentle lower slopes where the mountains meet the valley floor, carved out over millennia by 'La Grande Eau', surely a more romantic name in French than its English translation. The immediate visual picture for someone arriving for the first time in the valley is of gentle green pastures, beautiful meadows whose colours vary with the seasons, surrounded by the most fantastic sculptures of weathered stone, and deep forests, towering above. The alpine landscape always gives the impression of energy and vitality, with a clear blue sky broken by a line of peaks forming jagged angles. In winter, when the sun shines, the effect of the snow cover is vividly reflective and almost super luminary. The mood changes when the dark grey lowering clouds let fall a blanket of snow, covering the trees, chalets and paths. The Ormonts Valley has a length of 25 kilometres, born at the Col du Pillon, the pass to Gstaad, and spilling out onto the Rhone plain at Aigle, ten kilometres from the head of the lake at Villeneuve.

Even up to the nineteenth century, the mode of exploitation of the terrain was highly specialised, and developed to make best use of the local conditions. Land prone to avalanches in winter provided the best grazing in summer. Every family had different properties that they used according to the season. Gradually the ancestors cleared forest from the steep slopes for pasturage. Hay meadows of natural alpine grasslands were harvested in late summer. Vegetation in the Swiss Alps makes use of water from the permanent snow-line, saturating the ground in spring with turbid waters from melting glaciers that naturally irrigate the high pastures. Warmer continental summers encourage the lush alpine meadows to grow, providing the sweet grasses for the cattle, sheep and goats.

The clouds that form in the south-west, blown on the wind from the Atlantic, first encounter the Mont Blanc massif, where they dump much of the rain before crossing the Rhone plain to reach the Ormonts Valley. The necessity to work with the weather meant that the people became highly skilled forecasters, able to read the sky with great precision. The location in respect to the mountains, the colours, the patterns and types of cloud were all noted, memorised and transmitted down the generations.

This hidden valley was off the beaten track, far from the roads of Rome. The first historical mention is the gift of the territory to the Abbey of Saint Maurice in 515, albeit in a vague way as part of all the Alps from the head of the lake to Martigny, found in a document from 1400 by historian Corthesy. However, this gift may have excluded the Ormonts Valley, as there was no mention of the terrain below the Alps. The valley was thickly forested from the valley floor to the tree line, and avalanches fell at certain places which today are still called Lavanchy or Lavanches. Bears, wolves and lynxes roamed the slopes, and the chamois, marmottes and bouquetins occupied the highest terrain. The clearing of the valley walls for pasturage was a long, slow process that continued up to the end of the nineteenth century. Gradually, every settler cleared land, in amongst the forest reserved for timber, and used the pasturage around the family property to graze their livestock, and to cultivate the alpine meadows for winter hay. Settlers in the valley had to construct the first hut by hand, to shelter themselves and their animals. Forestry and wood-chopping skills were complemented by the use of fire for clearing. The glaciers above the valley at Les Diablerets descended much further than today, and this kept the humidity high in the valley, and the temperatures low.

All of the mountain villages in the valley grew from groups of chalets gathered together for mutual protection against the hostility and extremes of the environment, for example being snowed in for months at a time during harsh winters. The real mountain people were typical of those who live in difficult alpine terrain - tough, uncompromising survivors, often with a great store or courage and kindness, as well as being hardened by the elements. They became expert at reading the snow. Powder avalanches happen after the first falls of snow, when there is nothing to fix it to the hillside and where the slopes are so steep that the unstable powder slides off the mountainside and generally falls all the way to the bottom, resembling a cloud. The spring avalanches are caused by rain falling on the snow, and the sun's radiation warming the slopes, causing the slippage of built up layers of snow, of varying sizes, only coming to rest wherever the slope becomes a little less steep, halting the slide. However overall, the winter climate is invigorating, untroubled by the fog that blankets the Rhone plain and the lake. The air is pure, and the humidity from the snow keeps the temperature relatively mild, even almost warm when the sun shines from a cloudless sky. The valley is sheltered from the cold north winds.

The place names in the valley are very old, and an in depth analysis could help to date the first settlement. There are hundreds of terms and expressions that do not belong to the French language, in everyday usage until well into the twentieth century. They do not appear to have Germanic roots. The cadence and sound of the words reflect the ancient place names in the Upper Rhone Valley, and so probably are deeply linked to the dialects spoken locally prior to the formal adoption of Latin then French. They cover a complete range of activities and concerns that would have been used by pioneering settlers who built their own tools, houses, and developed unique methods of farming, animal husbandry and the names of plants and crops cultivated in the gardens and fields. This means that they are deeply rooted in the culture of the people who first settled the valley. These terms, preserved by local researchers from people who still understood the local patois, may provide a reflection of the language of the Nantuaten people who were the inhabitants at the time of the invasion by the army of ancient Rome. Perhaps further linguistic analysis could provide a link to other remnants of the ancient Gaulish language that gradually developed out of proto Indo European.

At the beginning of the twelfth century, the valley still belonged to the Abbey of Saint Maurice, which was shortly to lose all its territorial possessions to the Counts of Savoie. The feudal families that claimed possession of the region at this time were the Saillons, the Pontverres, and the de la Tours. A document from 1231 cedes St Triphon and the lower Ormont to Guy de Pontverre and the upper valley, covered in vast forests to the Saillons in 1287. According to Corthesy, the royal archives of Turin mentioned a 'William, chaplain of Ormont' in a charter postage of 1279 and a deed from 1287.

At this time there were settlements on the valley floor, and also at Le Sepey and La Forclaz. The Savoie families were successors to the Burgundians and Carolingians who had been despotic rulers of Helvetia during the dark ages. At this time our ancestral family, the Tardents, were settled at Le Sepey, as their name appears in documents from this period in the archives of Lausanne. A century or so later, and they had adopted Savoyard first names, and were chatelains, a kind of justice of the peace, able to represent the feudal families for local judicial and communal matters. Clearly one of their talents was adaptation to whatever situation in which they found themselves, and they took on a negotiation role between the local people and the Savoyards.

The Counts of Savoie used a line of sight to claim their feudal dominion, siting a castle in the Ormonts Valley at Aigremont. They then claimed that all the territory in view of the castle was theirs, and all the people who lived in the territory were their fiefs. After two centuries, Savoyard rule was a thing of the past. In 1233, Guy of Pontverre was cited as an administrator of Ormonts for the Abbey of Saint Maurice. In 1277, the Ormonts was referred to as 'Joria Hors Mont' a term used to describe the Pontverres claim to possession. Possibly 'hors monts' referred to the alpine pasturage above the valley floor that the Ormonans had hewn out of the forest during their occupation of the valley over the preceding centuries. The fact that they were claiming the valley meant that it was already well established, as if it was in its wilderness state, they would not be able to access the valley. Since the clearing of dense mountain forest takes generations, this is a clear indication that there was a local population who had been in residence for an indeterminate number of centuries.

From the beginning of the fourteenth century ownership of Ormonts was claimed by Guillaume of Pontverre and his nephew Aimone. Uncle and nephew pretended jurisdiction over the lands and fiefs of the Ormonts parish. In 1425, Antoine, Count of Gruyere, and Jean de la Baume, marshal of France divided up the estate of the Pontverres. Antoine left a natural son named Anthony, who sold the lordship of Aigremont to the Bernese for the price of 1,000 florins.

Tradition has it in the Ormonts Valley that the local people of Ormonts burned Aigremont castle after the Savoyards left. Another tradition provides that, during the time of the Pontverres, when there was a pressing danger from an invading army heading for Aigremont, the young people of Forclaz ran to warn a Pontverre lady in residence in the absence of her husband, Guy de Pontverre. The good lady, to show her gratitude, donated the great mountain of Perche, to the people of Forclaz, with the express condition that it be shared by women and men alike, and that girls from Forclaz who marry outside the valley were to maintain this right, granted for posterity. This tradition was still maintained in the 19th century, although the record of gift had long since vanished.

Out of the cradle of the melted Rhone Glacier, through the Celtic period of trade and transport over the Alpine passes, and after the settlement of the wild hidden Ormonts Valley, by no means was everyone involved in the gaining of riches and power through conquest. Most people, like our Swiss family Tardent, were locked into a quotidian existence, sometimes prospering, sometimes battling poverty. By the end of the 19th century the epoch of princely conquest was in its final phase, while a new militaristic expansionism in Europe took its place, and once again the influence of the struggle for wealth and power made itself felt in the more remote regions of Switzerland. For the traditional alpine farmers, still practising a form of transhumance for cattle, sheep and goats, land was in short supply for every new generation born into the mountains. Another mass migration from the old to the new world was the result, as once again people left behind deep roots, land, tradition, culture and language to seek new opportunities in fresh fields.

Sic transit gloria mundi.

BIBLIOGRAPHY

Books

The Ancient Paths: Discovering the Lost Map of Celtic Europe *Graham Robb*

Les Vignerons Suisses du Tsar *Olivier Grivat*

Le Chablais dans les pas des Archéologues *Carine Wagner*

Deep Simplicity: Chaos, Complexity and the Emergence of Life *John Gribbin*

Reinventing Gravity: A Physicist goes beyond Einstein *John Moffat*

Gaia: A New Look at Life on Earth *James Lovelock*

La Vallee des Ormonts *Paul Anex - Marie-Claude Busset-Henchoz -Henri-Louis Guignard - Edgar Pittex Denyse Raymond - Philippe Schoeneich*

Des Particularités du Parler aux Ormonts *Philippe Free, Marie-Claude Busset*

Papers

Le Tauredunum et la Géologie *Denis Fournier*

Climate Change - Environment and Civilization in the Middle East *A. Issar, Mattanyah Zohar*

Eurasian Steppes. Ecological Problems and Livelihoods in a Changing World *Marinus J.A. Werger, Marja A. van Staalduinen*

The Influence of Climatic Change on the Late Bronze Age Collapse and the Greek Dark Ages *Brandon Lee Drake*

Migration of Early Nomads of the Eurasian Steppe in the context of Climatic Change *N.A Bokovenko. The Institute for Material Culture, Russian Academy of Sciences.*

Impact of the Environment on Human Migration in Eurasia Proceedings of the NATO Advanced Research Workshop, held in St. Petersburg, 15-18 November 2003

Tracing the Origin and Spread of Agriculture in Europe *Ron Pinhasi, Joaquim Fort, and Albert J Ammerman*

Web Content

Letter from Henri Tardent to his descendants written at Nikolaiev Ukraine February - March 1887 *https://tardent-history.info/Letter_from_Henri.htm*

ABOUT THE AUTHOR

Nya Alison Murray

I decided to write about climate change because I think people deserve to know the facts about the precariousness of our situation. Urgent action is required, and we have to invest in renewable energy, clean technology and a sustainable biosphere now. So I started writing about our response to climate change.

It seems that the ancestors have been facing climate change, overuse of resources, and migration in search of opportunity for a long time, particularly since the interglacial period known as the Holocene.

I realize that we are all here today because of the resourcefulness and inventiveness of our ancestors, so I thought other people might be interested in our remarkable evolution through proto-history since the ice age. I also want to share my knowledge of the current position of us all, people, organizations, business, government and the UN in addressing the biggest challenge humanity has ever faced.

I am an Information Technology consultant and I am Australian. I surf, windsurf, and go snowboarding in the

Swiss Alps. I am an eco warrior, and love animals, plants, trees, the natural enivronment, beaches, the biosphere, the stars, the weather, the climate, global economics, music, dance and art, particularly Australian indigenous art, which I collect.

Made in the USA
Monee, IL
05 November 2026

67815700R00157